United States
Department of
Agriculture

Forest
Service

Agriculture
Handbook 674

March 2010

The Container Tree Nursery Manual

Volume Seven
Seedling Processing, Storage, and Outplanting

Thomas D. Landis, Retired National Nursery Specialist,
USDA Forest Service and Private Consultant, Medford, OR

R. Kasten Dumroese, National Nursery Specialist and Research Scientist,
USDA Forest Service, Rocky Mountain Research Station, Moscow, ID

Diane L. Haase, Western Nursery Specialist,
USDA Forest Service, State & Private Forestry, Portland, OR

Volume One   Nursery Planning, Development, and Management (1995)

Volume Two   Containers and Growing Media (1990)

Volume Three   Atmospheric Environment (1992)

Volume Four   Seedling Nutrition and Irrigation (1989)

Volume Five   The Biological Component: Nursery Pests and Mycorrhizae (1990)

Volume Six   Seedling Propagation (1999)

Volume Seven   Seedling Processing, Storage, and Outplanting (2010)

Landis, T.D.; Dumroese, R.K.; Haase, D.L. 2010.
The Container Tree Nursery Manual.
Volume 7, Seedling Processing, Storage, and Outplanting
Agric. Handbk. 674. Washington, DC: U.S. Department of Agriculture Forest Service. 200 p.

The use of trade or firm names in this publication is for reader information and does not imply endorsement by the Forest Service, U.S. Department of Agriculture of any product or service.

Pesticides used improperly can be injurious to humans, animals, and plants. Follow the directions and heed all precautions on the labels. Store pesticides in original containers under lock and key—out of the reach of children and animals—and away from food and feed. Apply pesticides so that they do not endanger humans, livestock, crops, beneficial insects, fish, and wildlife. Do not apply pesticides when there is danger of drift, when honey bees or other pollinating insects are visiting plants, or in ways that may contaminate water or leave illegal residues. Avoid prolonged inhalation of pesticide sprays or dusts; wear protective clothing and equipment if specified on the container. If your hands become contaminated with a pesticide, do not eat or drink until you have washed. In case a pesticide is swallowed or gets in the eyes, follow the first-aid treatment given on the label, and get prompt medical attention. If a pesticide is spilled on your skin or clothing, remove clothing immediately and wash skin thoroughly. Do not clean spray equipment or dump excess spray material near ponds, streams, or wells. Because it is difficult to remove all traces of herbicides from equipment, do not use the same equipment for insecticides or fungicides that you use for herbicides. Dispose of empty pesticide containers promptly. Have them buried at a sanitary land-fill dump, or crush and bury them in a level, isolated place. NOTE: Some States have restrictions on the use of certain pesticides. Check your State and local regulations. Also, because registrations of pesticides are under constant review by the Federal Environmental Protection Agency, consult your county agricultural agent or State extension specialist to be sure the intended use is still registered.

The U.S. Department of Agriculture (USDA) prohibits discrimination in all its programs and activities on the basis of race, color, national origin, age, disability, and where applicable, sex, marital status, familial status, parental status, religion, sexual orientation, genetic information, political beliefs, reprisal, or because all or part of an individual's income is derived from any public assistance program. (Not all prohibited bases apply to all programs.) Persons with disabilities who require alternative means for communication of program information (Braille, large print, audiotape, etc.) should contact USDA's TARGET Center at (202) 720-2600 (voice and TDD). To file a complaint of discrimination, write USDA, Director, Office of Civil Rights, 1400 Independence Avenue, S.W., Washington, D.C. 20250-9410, or call (800) 795-3272 (voice) or (202) 720-6382 (TDD). USDA is an equal opportunity provider and employer.

The Container Tree Nursery Manual

Volume Seven
Seedling Processing, Storage, and Outplanting

Chapter 1—The Target Plant Concept
- 7.1.1 Introduction *3*
- 7.1.2 Defining the Target Plant *4*
- 7.1.3 Field Testing the Target Plant *11*
- 7.1.4 Summary *13*
- 7.1.5 Literature Cited *14*

Chapter 2—Assessing Plant Quality
- 7.2.1 Introduction *19*
- 7.2.2 Classes of Plant Quality Attributes *20*
- 7.2.3 Morphological Attributes *21*
- 7.2.4 Physiological Attributes *27*
- 7.2.5 Performance Attributes *52*
- 7.2.6 Correlating Combinations of Plant Quality Tests To Predict Outplanting Performance *68*
- 7.2.7 Limitations of Plant Quality Tests *69*
- 7.2.8 Commercial Plant Quality Testing Laboratories *71*
- 7.2.9 Summary and Conclusions *72*
- 7.2.10 Literature Cited *74*
- 7.2.11 Appendix *81*

Chapter 3—Harvesting
- 7.3.1 Introduction *85*
- 7.3.2 Scheduling the Winter Harvesting Window *87*
- 7.3.3 Pre-storage Fungicide Treatments *90*
- 7.3.4 Processing Speculation and Contract *91*
- 7.3.5 Grading and Packaging *92*
- 7.3.6 Packaging for Storage and Shipping *99*
- 7.3.7 Processing Cull Seedlings *100*
- 7.3.8 Summary and Conclusions *101*
- 7.3.9 Literature Cited *102*

Chapter 4—Plant Storage
- 7.4.1 Introduction *107*
- 7.4.2 Short-Term Storage for Summer or Fall Outplanting—"Hot-Planting" *108*
- 7.4.3 Overwinter Storage *110*
- 7.4.4 Nonrefrigerated Storage Systems *112*
- 7.4.5 Refrigerated Storage *120*
- 7.4.6 Monitoring Plant Quality in Storage *126*
- 7.4.7 Causes of Overwinter Damage *128*
- 7.4.8 Summary and Conclusions *131*
- 7.4.9 Literature Cited *132*

Chapter 5—Handling and Shipping
- 7.5.1 Introduction *137*
- 7.5.2 Minimizing Stresses During Handling *138*
- 7.5.3 Handling and Shipping Systems *142*
- 7.5.4 Nursery Stock Delivery *145*
- 7.5.5 Summary and Recommendations *148*
- 7.5.6 Literature Cited *149*

Chapter 6—Outplanting
- 7.6.1 Introduction *154*
- 7.6.2 Outplanting Windows *155*
- 7.6.3 Onsite Handling and Storage *157*
- 7.6.4 Pre-Planting Preparations *162*
- 7.6.5 Selecting Plant Spacing and Pattern *169*
- 7.6.6 Crew Training and Supervision *171*
- 7.6.7 Hand-Planting Equipment *173*
- 7.6.8 Machine Planting *179*
- 7.6.9 Planting Equipment for Large Stock *184*
- 7.6.10 Treatments at Time of Planting *186*
- 7.6.11 Monitoring Outplanting Performance *190*
- 7.6.12 Conclusions and Recommendations *193*
- 7.6.13 Literature Cited *194*

Preface

The Container Tree Nursery Manual consists of seven volumes that have all been published under the same series number: USDA Agriculture Handbook 674. Writing began in the late 1980s, with the first volume published in 1989. Subsequent volumes were published at increasingly longer intervals with the seventh and last volume taking over 10 years to complete (fig. 1).

Each volume contains chapters on closely related subjects concerning the production of trees and other native plants in containers. The volumes can be accumulated and used as a complete nursery manual, or they can be used separately by specialists needing information on a particular subject. Because several subjects must be discussed in more than one volume, there will be some redundancy in the manual. Such repetition is justified, however, because most readers will be using the manual as a technical reference and will not be reading the entire text.

The Container Tree Nursery Manual has been functionally organized to follow the normal sequence of nursery development, seedling propagation, and outplanting. Volume one discusses the various steps that should be followed in developing a nursery facility. Volume two is concerned with the selection of types of containers and growing media. Volume three and volume four analyze the "limiting factors" that affect seedling growth and discuss how they can be manipulated in container nurseries. Volume five examines the various biological organisms that can affect seedlings, either negatively as pests or positively as mycorrhizae. Volume six shows how to develop growing schedules and how seedlings are propagated through the three growth phases. Volume seven covers the time from when the crop is hardened-off and ready for harvest to when they go in the ground.

The seven volumes are structured around an outline of numerical organizational headings that enable the reader to locate a specific subject quickly, without referring to an index. The general outline of volume and chapter titles is as follows:

Volume One—Nursery Planning, Development, and Management
Chapter 1 Initial Planning and Feasibility Assessment
Chapter 2 Site Selection
Chapter 3 Nursery Design and Site Layout
Chapter 4 Environmental Control and Safety Equipment
Chapter 5 Service Buildings and Equipment
Chapter 6 Nursery Management
Chapter 7 Troubleshooting Nursery Problems

Volume Two—Containers and Growing Media
Chapter 1 Containers: Types and Functions
Chapter 2 Growing Media

Volume Three—Atmospheric Environment
Chapter 1 Temperature
Chapter 2 Humidity
Chapter 3 Light
Chapter 4 Carbon Dioxide

Volume Four—Seedling Nutrition and Irrigation
Chapter 1 Mineral Nutrition and Fertilization
Chapter 2 Irrigation and Water Management

Volume Five—The Biological Component: Nursery Pests and Mycorrhizae
Chapter 1 Disease and Pest Management
Chapter 2 Mycorrhizae

Volume Six—Seedling Propagation
Chapter 1 Crop Planning
Chapter 2 Seed Propagation
Chapter 3 Vegetative Propagation
Chapter 4 Seedling Development: The Establishment, Rapid Growth, and Hardening Phases

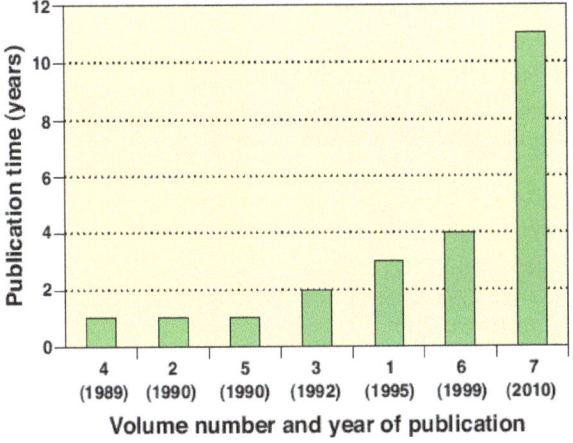

Figure 1—*Years required to write each volume of Agriculture Handbook 674.*

Volume Seven—Seedling Processing, Storage, and Outplanting
Chapter 1   The Target Plant Concept
Chapter 2   Assessing Plant Quality
Chapter 3   Harvesting
Chapter 4   Plant Storage
Chapter 5   Handling and Shipping
Chapter 6   Outplanting

This manual is based on the best current knowledge of container nursery management and should be used as a general reference. Recommendations were made using the best information available at the time and are, therefore, subject to revision as more knowledge becomes available. Much of the information in this manual was primarily developed from information on growing western and southern conifer seedlings in the United States. Because of the wide variation in individual species responses, container nursery managers will need to adapt these principles and procedures to their own crop requirements. There is no substitute for individual experience, and recommended cultural practices should be tested before being implemented on an operational scale.

Trade names are used throughout the manual, but only to provide examples, and no endorsement by USDA of specific products, or exclusion of equally suitable products, is implied. The mention of specific pesticides is intended only for general information and should not be construed as an endorsement. Because of frequent changes in pesticide registration and labeling, the reader should check with local authorities to make sure that an intended use is both safe and legal. Remember that pesticides can be harmful to humans, domestic animals, desirable plants, and fish or other wildlife if they are not handled or applied properly. Use all pesticides selectively and carefully, following the label directions. Follow recommended practices for the disposal of surplus pesticides and pesticide containers.

Acknowledgments
Many people have been instrumental in the preparation of the manual. Amy Grey and Jim Marin were responsible for layout and production.

Technical review of such a large publication involves considerable work, and the authors thank the following nursery professionals for reviewing final drafts of this volume:

All six Chapters:
    John Mexal
    Steve Grossnickle
    Nabil Khadduri
    Doug McCreary
Chapter 1 - The Target Plant Concept
    Douglass Jacobs
    David South
    Glenda Scott
Chapter 2 - Assessing Plant Quality
    David South
    Conor O' Reilly
Chapter 4 - Plant Storage
    David G. Simpson
Chapter 6 - Outplanting
    Glenda Scott
    Leo Tervo
    Risto Rikala

Where To Obtain Copies
The Forest Service initially purchased a limited quantity of hard copies of each volume for free distribution but many of the earlier volumes are out of print. Because of the cost of reprinting in color, Stuewe and Sons has reprinted volumes in black and white. All of the volumes have also been published as electronic books ("e-books") in Adobe PDF format. Both hard copies and e-books can be purchased from either of the following sources; contact them for current availability and prices.

Western Forestry and Conservation Association
4033 SW Canyon Road
Portland, OR 97221 USA
Tel: 503-226-4562 Fax: 503-226-2515
E-mail: richard@westernforestry.org
Web site: http://www.westernforestry.org

Stuewe & Sons, Inc.
31933 Rolland Drive
Tangent, OR 97389 USA
Tel: 1-800-553-5331 or 541-757-7798
Fax: 541-754-6617
E-mail: info@stuewe.com
Web site: http://www.stuewe.com

In addition, PDFs of each volume may be viewed and downloaded from the Reforestation, Nurseries, and Genetics Resources Web site: http://rngr.net.

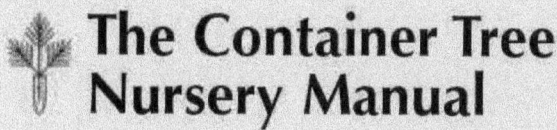
The Container Tree Nursery Manual

Volume Seven

Chapter 1
The Target Plant Concept

Contents

7.1.1 Introduction 3

7.1.2 Defining the Target Plant 4
7.1.2.1 Objectives of the outplanting project 4
7.1.2.2 Type of plant material 4
    Species
    Stocktype
7.1.2.3 Genetic considerations 6
    Local adaptation
    Genetic diversity
    Sexual diversity
7.1.2.4 Limiting factors on the outplanting site 7
7.1.2.5 Timing of the outplanting window 8
7.1.2.6 Outplanting tools and techniques 10

7.1.3 Field Testing the Target Plant 11

7.1.4 Summary 13

7.1.5 Literature Cited 14

7.1.1 Introduction

The basic ideas behind the **Target Plant Concept** can be traced back to the late 1970s and early 1980s when new insights into seedling physiology were radically changing nursery management. Forestry researchers began analyzing the effects of nursery cultural practices on outplanting performance and, as a consequence, foresters gave more thought to their reforestation prescriptions and began asking for new and different stocktypes (fig. 7.1.1). By 1990, the term *target plant* had become well established in nursery and reforestation jargon. In that year, the Target Seedling Symposium brought together foresters and nursery workers to discuss all aspects of the target plant, and the resultant proceedings are still a major source of information on the subject (Rose and others 1990).

One basic tenet of the Target Plant Concept is that plant quality is determined by outplanting performance (Landis 2002). Although they might be the same species, forest and conservation plants are very different from ornamental nursery stock. For example, Douglas-fir (*Psuedotsuga menziesii*) seedlings outplanted in relatively harsh forest environments will have different requirements from those outplanted in city parks or Christmas tree plantations. These differences are pivotal to the Target Plant Concept because plant quality depends on how the plants will be used—"fitness for purpose" (Sutton 1980). This means that plant quality cannot be merely described at the nursery; it must also be proven on the outplanting site. There is no such thing as an "all-purpose" plant because nice-looking plants at the nursery will not survive and grow well on all sites.

When defining a target plant for a particular project, economics and management objectives must be also considered. When different size classes of slash pine (*Pinus elliottii* var. *elliottii*) were outplanted and then measured after 4 years, seedlings with larger stem diameters had better survival and growth than the standard "shippable" nursery stock. An economic analysis proved that these larger plants were the best investment (South and Mitchell 1999).

Figure 7.1.1—*The "Target Plant Concept" developed as foresters and other plant users began to work more closely with nurseries to develop stocktypes for specific outplanting projects.*

7.1.2 Defining the Target Plant

A target plant is one that has been cultured to survive and grow on a specific outplanting site, and that can be defined in six sequential components (fig 7.1.2).

7.1.2.1 Objectives of the outplanting project

The reasons why nursery stock is needed will have a critical influence on the characteristics of the target plant. In traditional reforestation, a commercially valuable tree species that has been genetically improved for fast growth, good form, or desirable wood quality may be outplanted with the ultimate objective of producing saw logs or pulp.

The target plant for a restoration project, however, might be radically different because the objectives are totally different. For example, a watershed protection project would require riparian trees and shrubs and wetland plants that will not be harvested for any commercial product. In this case, the objectives would include stopping erosion, stabilizing the stream bank, and ultimately restoring a functional plant community. Fire restoration projects will have different objectives depending on the plant community type and the ultimate use of the land. Project objectives for a burned rangeland might be to stop soil erosion, replace exotic weed species with native plants, and establish browse plants for deer or elk. Target plants for such a project might include a direct seeding of native grass and forbs, followed by an outplanting of woody shrub nursery stock. For a burned forest, however, the plant materials might be native grass seeds to stop erosion and then outplanting of tree seedlings to bring the land back to full productivity as soon as possible. Another project might be to restore plants that are in danger of going extinct in a particular habitat. For example, Short's goldenrod (*Solidago shortii*) is an endangered plant that can be found only in 14 populations in a small geographic area in Kentucky (Baskin and others 2000). Fortunately, this plant is relatively easy to propagate from seeds and grows well in greenhouses.

Conservation planting projects can have still different objectives. Although native plants are emphasized whenever and wherever possible, exotic species may be required on extreme sites. In dry areas of the Intermountain West, where no native trees for upland sites are available, species such as Austrian pine (*Pinus nigra*) and Siberian elm (*Ulmus pumila*) are used to create windbreaks for home or livestock protection. Project objectives are a critical first consideration in the target plant concept.

7.1.2.2 Type of plant material

The second consideration in the Target Plant Concept is what types of plant material would be best (fig. 7.1.2). Plant materials refer to anything that can be used to propagate a species; these propagules can be seeds, bulbs or rhizomes, cuttings, or seedlings (Landis 2001). In container nurseries, plant material usually means the species and the stocktype.

Species. The species is determined by the project objectives that were discussed in the previous section. For example, Douglas-fir is one of the most important timber species in the Pacific Northwest and is therefore a major crop in local forest nurseries. Douglas-fir has been outplanted extensively for the past century, often in monocultures. In coastal areas of Oregon and Washington, these pure stands have recently become severely infected with Swiss needle cast caused by the fungus *Phaeocryptopus gaeumannii*. One silvicultural recommendation is to interplant with other conifers, especially western hemlock (*Tsuga heterophylla*), to reduce the impact of this disease (Filip and others 2000). In the Southeastern United States, the demand for longleaf pine (*Pinus palustris*) has increased tremendously in recent years and, for this species, container stock has proven to survive and grow better than bareroot stock (Barnett 2002).

Stock type. Container nurseries are currently produce a wide variety of stocktypes, including seedlings, transplants, and rooted cuttings. Although biological factors should be the primary consideration, the choice of container stocktype is

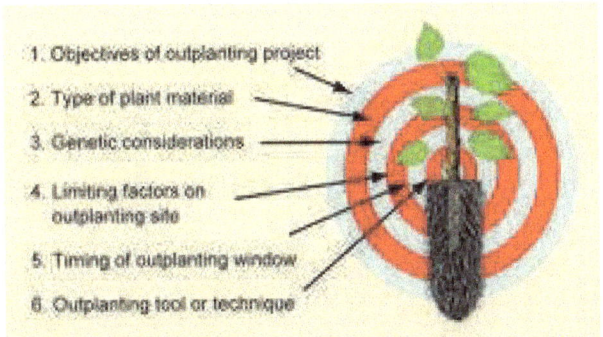

Figure 7.1.2—*The six components of the Target Plant Concept.*

primarily defined by price and preference. Experienced nursery customers consider the cost per surviving plant when deciding on stocktype and other target seedling factors.

Selling Price—Although the cost of containers and growing media are important, the price of container stock is basically a function of nursery production space. A unit area of greenhouse bench space costs a fixed amount, so the prices of the various container sizes increase as their cell densities decrease (table 7.1.1). Actual selling prices for each container size are set by market factors, especially demand and effects of competition.

Table 7.1.1—Container seedling selling price is primarily a function of nursery production space

Type of container		Cell volume		Number of cells per		Price per
		cm³	in³	m²	ft²	1,000 seedlings ($)*
Styroblock™ 1	207A	8	1.1	2,121	196	100
Styroblock™ 2A	211A	41	2.5	1,032	103	190
Styroblock™ 5.5	315B	90	5.5	756	71	276
Styroblock™ 10	415D	160	9.8	364	34	576
Styroblock™ 15	515A	250	15.3	284	26	755
Styroblock™ 20	615A	336	20.5	213	20	980

* Arbitrarily set price, U.S. dollars, 2007.

Figure 7.1.3—Larger container plants are gaining in popularity (A), but outplanting trials are needed to determine which sizes grow best and are most economical. Eight years after outplanting, spruce seedlings in containers that were 340 cm³ (20 in³) in volume were the best choice on sites with heavy vegetative competition in Quebec (B).

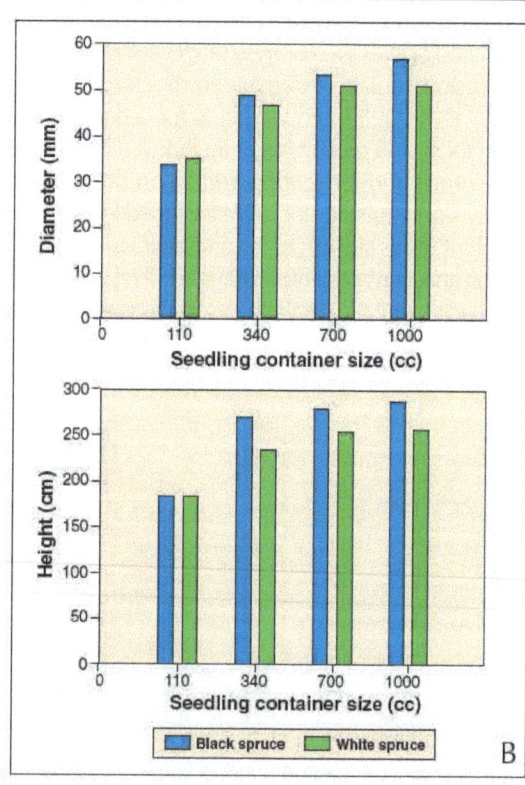

Customer Preference—The demand for container types has changed considerably over the past 25 years, and one trend is to larger volumes. For example, in the 1970s, one Oregon nursery typically produced container stock of 33 to 66 cm^3 (2 to 4 in^3), whereas, by year 2000 they were growing all their seedlings in 246 to 328 cm^3 (15 to 20 in^3) containers (fig. 7.1.3A). This preference for larger stocktypes has led to the practice of container transplanting, where seedlings are started in small "miniplugs" in greenhouses and then transplanted to larger containers grown in outdoor compounds.

One reason for larger container stocktypes to be in greater demand is because of increased vegetative competition on the outplanting site. Other factors being equal, plants grown in larger containers have larger caliper and a better shoot-to-root ratio, which gives them an advantage on sites with heavy competition. Environmental concerns in Quebec have led to a prohibition of herbicide use for site preparation. The standard stock size for black spruce (*Picea mariana*) and white spruce (*Picea glauca*) on these sites was 110 cm^3 (7 in^3), and, therefore, research trials were established to test a range of larger container sizes (Jobidon and others 2003). When measured 8 years after outplanting (fig. 7.1.3B), seedlings in the 340 cm^3 (20 in^3) containers were found to be the best and most economical stocktype in the absence of herbicides.

Customer preferences are also evidenced by regional trends in container type. It is cost prohibitive for a nursery to test all types of containers, so they typically use whatever is locally popular. Styroblock™ containers were developed in British Columbia and continue to be the most popular container type in the Pacific Northwest (Van Eerden 2002). In the Northeastern United States and Canada, however, hard plastic Ropak® Multi-Pots were the most popular container type and now are being replaced by Jiffy® cells (White 2003).

7.1.2.3 Genetic considerations

The third consideration of the Target Plant Concept concerns the question of genetics. Three factors should be considered: local adaptation, genetic diversity, and sexual diversity.

Local adaptation. Many native plants can be propagated by seeds collected on or near the project area. "Seed source" is an idea familiar to all forest nursery managers and reforestation specialists who know that, because plants are adapted to local conditions, seeds should always be collected within the local "seed zone." Container nurseries grow plants by seed zone, which is a three-dimensional geographic area that is relatively similar in climate and soil type (see Volume Six, Section 6.2.1.2). Local adaptation is not always considered in ornamental nurseries. For example, both native plant nurseries and ornamental nurseries grow Douglas-fir seedlings but the former distinguish between ecotypes (for example, variety *glauca*) and ornamental nurseries offer different cultivars (for example, 'Carneflix Weeping') (Landis 2001).

Seed source affects plant performance in several ways, especially growth rate and cold tolerance. In general, plants grown from seeds collected from higher latitudes or elevations will grow more slowly and tend to be more cold hardy during winter than those grown from seeds collected from lower elevations or more southern latitudes (St. Clair and Johnson 2003). Seed zone research has not been done on many native plants, but it is intuitive that the same concepts should apply. Therefore, it would be prudent to always collect seeds or cuttings from the same geographic zone and elevation in which the nursery stock is to be outplanted. With the increasing concern about global climate change, there are likely to be adjustments in seed transfer guidelines with the strategic goal of encouraging gradual adaptation based on the latest research (Millar and others 2007).

Genetic diversity. Target plants should also represent the genetic diversity present on the outplanting site. Again, future climate change should be considered, especially for long-lived tree species. To maximize genetic diversity in the resultant seedlings, seeds should be collected from as many different plants as possible. The same principles apply to plants that must be propagated vegetatively. Cuttings must be collected near the outplanting site to make sure they are properly adapted. Of course, collecting costs must be kept within reason, so the number of seeds or cuttings collected must be a compromise. Guinon (1993) provides an excellent discussion of all factors involved in preserving biodiversity when collecting seeds or cuttings, and suggests collecting from at least 50 to 100 donor plants.

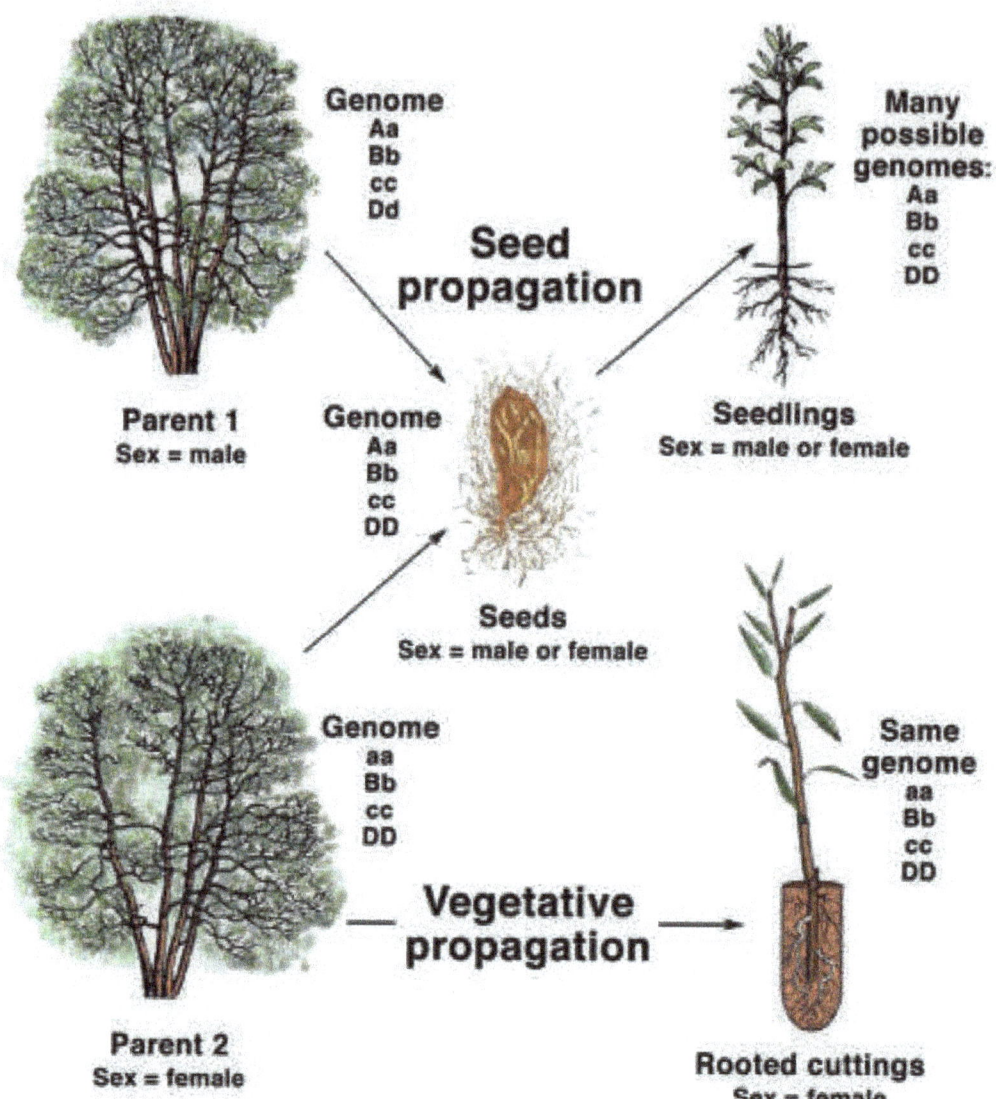

Figure 7.1.4—*The choice of whether to propagate by seeds or cuttings will affect the genetic diversity of the resultant crop. With dioecious plants, such as willows and cottonwoods, the sex of the parent plant must also be considered to make sure that the outplanting contains a mixture of both males and females (modified from Landis and others 2003).*

Sexual diversity. Dioecious plants, such as *Salix* and *Populus*, present another consideration, because all progeny produced by vegetative propagation will have the same sex as their parent (fig. 7.1.4). Therefore, when collecting cuttings at the project site, care must be taken to ensure that both male and female plants are approximately represented. Willows, cottonwoods, and aspen are sexually precocious so another option is to collect sexually mature cuttings from a broad genetic base that represents both sexes and root them in a nursery. Within 1 to 2 years the cuttings will flower and produce seeds. The seeds can then be sown into containers and the resultant seedlings will have a broad genetic and sexual diversity (Landis and others 2003).

7.1.2.4 Limiting factors on the outplanting site

The fourth consideration of the Target Plant Concept is based on the ecological "principle of limiting factors," which states that any biological process will be limited by that factor present in the least amount. Each outplanting site should be evaluated to identify the environmental factors most limiting to survival and growth (fig. 7.1.5A). Foresters do this when they write prescriptions for each harvest unit, specifying which tree species and stocktype would be most appropriate (fig. 7.1.1).

On most reforestation sites, soil moisture is the limiting factor and target plant specifications often reflect this fact.

At northern latitudes or at high elevation, however, cold soil temperatures may be more significant than soil moisture. Access to these sites may be restricted by snow that may not melt until late June or even July (Faliszewski 1998; Fredrickson 2003). The melting snow keeps soil temperatures cool and this can be limiting as research has shown that plant root growth is restricted below 10 °C (50 °F) (fig. 7.1.5B) (Lopushinsky and Max 1990). A reasonable target plant for these sites could be grown in a relatively short container to take advantage of warm moist surface soils (fig. 7.1.5C) (Landis 1999), as is the case for high elevation reforestation sites in British Columbia (Faliszewski 1998).

Restoration sites pose interesting challenges when evaluating outplanting sites for limiting factors. For example, after a wildfire, soil conditions are often severely altered, whereas mining sites have extreme soil pH levels. Riparian restoration projects require bioengineering structures to stabilize streambanks and retard soil erosion before the site can be planted (Hoag and Landis 2001). In desert restoration, low soil moisture, hot temperatures, high winds with sand blast, and heavy grazing have been listed as limiting factors (Bainbridge and others 1992).

Animal predation and snow load can also be limiting factors on some outplanting sites, especially at high elevations in the mountains. Container Engelmann spruce (*Picea engelmannii*) seedlings of various diameter grades were outplanted on a mountainous site in northern Utah. After two seasons, seedlings with larger diameters had significantly higher survival than those with smaller ones. Stock with larger diameters showed less mortality from snow breakage or rodent depredation (Hines and Long 1986).

One potential limiting factor that deserves special consideration: mycorrhizal fungi. These symbiotic organisms provide their host plants with many benefits, including better water and mineral nutrient uptake. Reforestation sites typically have an adequate complement of mycorrhizal fungi that quickly colonize outplanted nursery stock, whereas many restoration sites do not. For example, severe forest fires or surface mining eliminate all soil microorganisms, including mycorrhizal fungi. Therefore, plants destined for such sites should be inoculated with the appropriate fungal symbiont before outplanting. (See Volume Five, Chapter 2, for a complete discussion of mycorrhizae).

These examples demonstrate why nursery managers must work closely with plant customers to identify which environmental factors will be most limiting on each outplanting site. Through such discussions, specifications for the best target plant material can be designed to maximize survival and growth under specific site conditions.

7.1.2.5 Timing of the outplanting window

The outplanting window is the period of time in which environmental conditions on the outplanting site are most favorable for survival and growth of seedlings or rooted cuttings. The outplanting window is usually defined by limiting factors and, as discussed in the previous section, soil moisture and temperature are the usual constraints. In most of the continental United States and Canada, nursery stock is outplanted during the rains of winter or early spring when soil moisture is high and evapotranspirational losses are low (fig. 7.1.6). Obviously, the specific dates of the winter outplanting window will change with latitude and elevation, being earlier in the south and at low elevations and later farther north and at higher elevations.

One important advantage of container plants is that they can be sown at different dates and then cultured to be physiologically conditioned for outplanting during different times of the year. For the traditional outplanting windows of winter or early spring, plants can be harvested and hot-planted or cooler-stored for a few weeks until the outplanting site is ready (fig. 7.1.7A). As mentioned in the previous section, high elevation or boreal sites are challenging because they cannot be accessed during the typical midwinter outplanting window. Outplanting during the fall has been tried for decades with varying results. In recent years, however, interest in fall outplanting has been renewed, which is primarily due to the availability of properly conditioned container stock (Fredrickson 2003). In the Southeastern United States, the traditional outplanting window for loblolly pine is during the winter but container stock can be outplanted in the fall if hardened with shortened photoperiod in a greenhouse or exposed to naturally cooler temperatures in an outdoor compound for 6 weeks (Mexal and others 1979).

Summer outplanting is a relatively new practice that developed in the boreal regions of Canada (Revel and others 1990) and has since found some application at high elevation sites in the Rocky Mountains (Scott 2006).

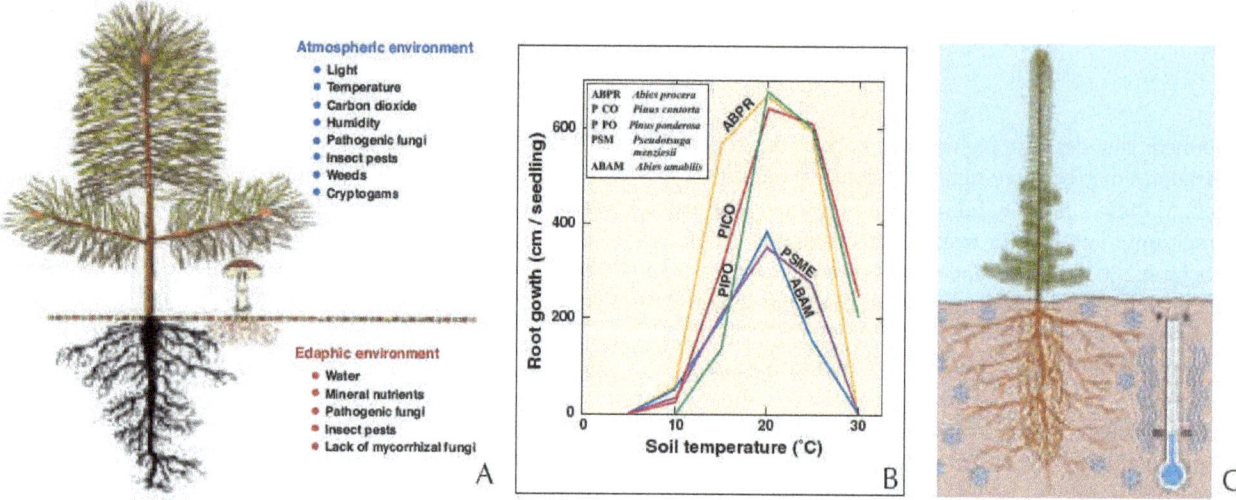

Figure 7.1.5—*A key part of the Target Plant Concept is to evaluate which environmental factors may be limiting on the outplanting site (A). At high elevations and latitudes, spring soil temperatures are cold and research has shown that roots of many commercial conifers do not grow appreciably below 10 °C (50 °F)(B). Therefore, target plants for these sites should have a relatively short, compact root system to take advantage of the warmer temperatures in the surface soil layers (C) (B, modified from Lopushinsky and Max 1990).*

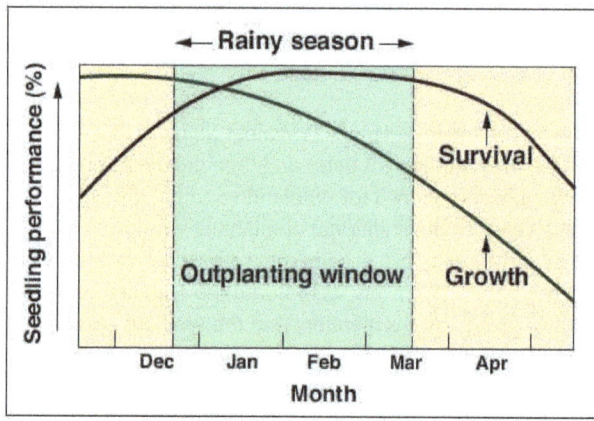

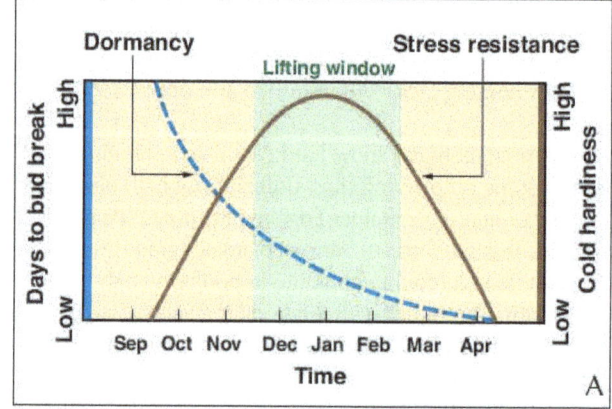

Figure 7.1.6—*A critical component of the Target Plant Concept is the "outplanting window," which is defined as the period of time in which plant survival and growth are optimal for that particular site. In much of the United States, the outplanting window is during the rainy period of midwinter (modified from South and Mexal 1984).*

Figure 7.1.7—*Container plants can be grown to meet the target requirements for a variety of outplanting windows. They can be harvested at their peak of physiological quality for the traditional midwinter window (A), or be specially cultured for summer or fall outplanting (B).*

Target plant characteristics are significantly different for spring versus summer or fall outplanting (Grossnickle and Folk 2003). Because they are less cold hardy and stress resistant, plants for summer and fall outplanting must be handled more carefully during shipping and onsite storage.

7.1.2.6 Outplanting tools and techniques

Each outplanting site has an appropriate tool; therefore tools and outplanting techniques must be considered in the Target Plant Concept. All too often, foresters or restoration specialists develop a preference for a particular implement because it has worked well in the past. However, no one tool will work under all site conditions. Although outplanting tools are discussed in detail in Chapter 7.6, a couple of examples of how outplanting tools and techniques can affect target plant specifications are mentioned here.

Soon after development of the first container plants, special implements were designed to outplant them (Hallman 1993). Dibbles were constructed in the exact same size and shape as the container plugs and the Pottiputki was designed to plant paperpot plants (fig. 7.1.8A). Nursery stock that is outplanted mechanically imposes unique restrictions because the target plant must conform to the size and shape of the handling equipment. Plants used in machine-powered planting equipment must have stem diameters that fit the holding clips, and root systems must not be longer than the depth of the furrow. The newest and most sophisticated machine-powered planting equipment requires plants of a size and shape that can be pneumatically loaded into planting heads (fig. 7.1.8B). So, where mechanical planting is used, the size and shape of the target plant must match the type of outplanting tool as well as biological conditions on the outplanting site.

New outplanting tools are continually being developed. Specially modified hoedads called "plug hoes" are now available for container stock. Again, nursery managers must work closely with reforestation or restoration project managers to make certain that their target plants can be properly outplanted in the soil conditions on the project site. The "tall pots" used in many restoration projects require specialized outplanting equipment. The Expanded Stinger uses an articulated planting head to place tall-pot seedlings or cuttings in compacted soil or even rock (Steinfeld and others 2002) (fig. 7.1.8C).

Figure 7.1.8—*The type of outplanting tool has a significant effect on the target plant. Hand-planting tools, such as the Pottiputki (A), were developed to handle paperpot seedlings, one specific type of container stock. With mechanical planting machines (B), plants must be grown in a particular size and shape to fit the handling system. The special stocktypes needed for restoration projects require innovative new outplanting equipment, such as the Expanded Stinger, which was developed for tall pots (C).*

7.1.3 Field Testing the Target Plant

Properly applied, the Target Plant Concept is a collaboration among nursery managers and their customers. At the start of any planting project, the customer and the nursery manager should agree on certain morphological and physiological specifications. This prototype target plant is grown in the nursery and then verified by outplanting trials that monitor survival and growth for up to 5 years (fig. 7.1.9).

Monitoring plant survival and growth during the first few months after outplanting is critical because problems with stock quality show up soon after outplanting. Problems with poor planting or exposure to drought conditions take longer to appear; plants exhibit good initial survival but gradually lose vigor and perhaps die. Therefore, plots must be monitored during the first month or two after outplanting and again at the end of the first year for initial survival. Subsequent checks after 3 to 5 years will give a good indication of plant growth rates. This performance information is then used to give valuable feedback to the nursery manager, who can fine-tune the target specifications for the next crop.

For example, the Oregon State University Nursery Technology Cooperative is conducting outplanting trials of 1-year-old stocktypes on two fire restoration sites in southwestern Oregon (Nursery Technology Cooperative 2005). The Timbered Rock site in the Cascade Mountains is much drier than the Biscuit site in the Coast Range. In terms of survival, the Styroblock™ container performed much better than the transplants at Timber Rock, whereas little difference was noted on the wetter Biscuit site (table 7.1.2). The container stocktype also grew much better at both sites, but especially so at Timbered Rock where grass competition was severe. In fact, the severe moisture stress caused by the grass resulted in a negative stem growth for the two transplant stocktypes. After 3 years, however, the container stocktype exhibited severe chlorosis and slower growth rates, which demonstrates the need for repeated monitoring to accurately assess seedling and stocktype performance.

Figure 7.1.9—*The target plant is not a fixed concept, but rather must be continually updated with information from outplanting trials.*

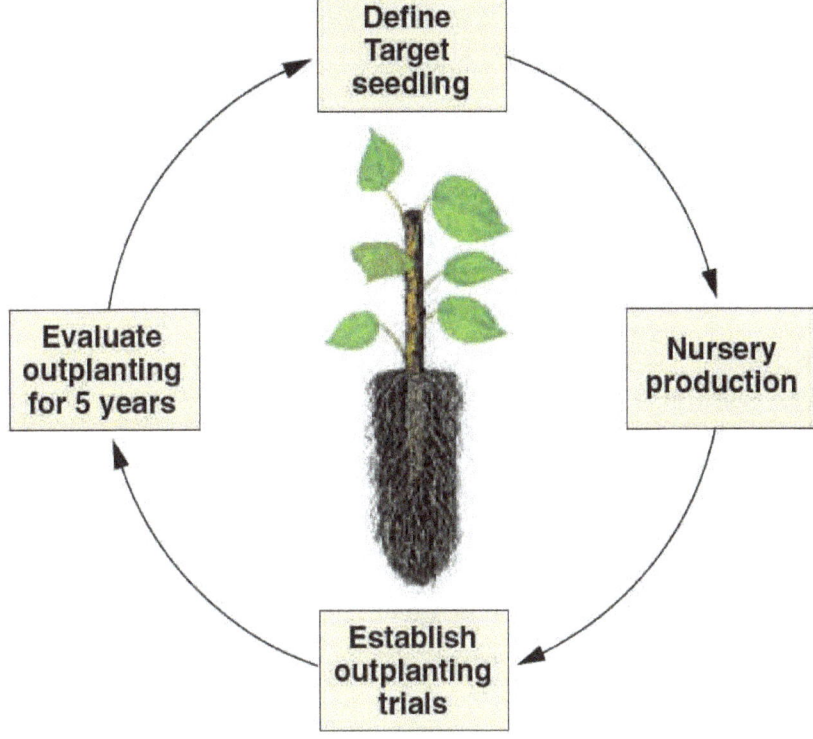

Table 7.1.2—Outplanting performance of Douglas-fir stocktypes on different outplanting sites after one growing season

Stocktype	Survival (%)	Height growth (cm)	Stem diameter growth (mm)
Timbered Rock Fire— Oregon Cascade Mountains			
1+1 bareroot transplant	14 c*	4.2 b	– 0.6 b
Q-plug container transplant	39 b	2.6 b	– 0.3 b
Styroblock™ container (246 cm³)	87 a	12.0 a	0.8 a
Biscuit Fire— Oregon Coastal Mountains			
1+1 bareroot transplant	98 a	4.6 b	0.5 b
Q-plug container transplant	98 a	7.0 a	0.5 b
Styroblock™ container (246 cm³)	99 a	7.5 a	1.1 a

* Different letters in each column represent statistical differences at the $P = 0.05$ level.

7.1.4 Summary

The Target Plant Concept is a relatively new but effective way of looking at reforestation and restoration. It emphasizes that plant quality must be defined on the outplanting site, and that there is not one universal best stocktype. In particular, the Target Plant Concept emphasizes that successful outplanting projects require good communication between the plant user and the nursery manager. The Target Plant Concept should be viewed as a circular feedback system in which information from the outplanting site is used to define and refine the best type of plant for each project. Practical considerations for implementing a nursery program based on the Target Plant Concept can be found in Rose and Haase (1995).

7.1.5 Literature Cited

Bainbridge, D.A.; Sorensen, N.; Virginia, R.A. 1992. Revegetating desert plant communities. In: Landis, T.D., ed. Proceedings, Western Forest Nursery Association. Gen. Tech. Rep. RM-221. Fort Collins, CO: USDA Forest Service, Rocky Mountain Forest and Range Experiment Station: 21-26.

Barnett, J.P. 2002. Longleaf pine: Why plant it? Why use containers? In: Barnett, J.P.; Dumroese, R.K.; Moorhead, D.J., eds. Proceedings of workshops on growing longleaf pine in containers —1999 and 2001. Gen. Tech. Rep. SRS-56. Asheville, NC: USDA Forest Service, Southern Research Station: 5-7.

Baskin, J.M.; Walck, J.L.; Baskin, C.C.; Buchele, D.E. 2000. *Solidago shortii* (Asteraceae). Native Plants Journal 1(1): 35-41.

Faliszewski, M. 1998. Stock type selection for high elevation (ESSF) planting. Kooistra, C.M., ed. Proceedings of the 1995, 1996, 1997 Forest Nursery Association of British Columbia. Vernon, BC, Canada: BC Ministry of Forests, Nursery Services Office—South Zone: 152.

Filip, G.; Kanaskie, A.; Kavanagh, K.; Johnson, G.; Johnson, R.; Maguire, G. 2000. Silviculture and Swiss needle cast: research and recommendations. Research Contribution 30. Corvallis, OR: Oregon State University, College of Forestry. 16 p.

Fredrickson, E. 2003. Fall planting in northern California. In: Riley, L.E.; Dumroese, R.K.; Landis, T.D., tech. coords. National Proceedings, Forest and Conservation Nursery Associations—2002. Proceedings RMRS-P-28. Ogden, UT: USDA Forest Service, Rocky Mountain Research Station: 159-161.

Grossnickle, S.C.; Folk, R.S. 2003. Spring versus summer spruce stocktypes of western Canada: nursery development and field performance. Western Journal of Applied Forestry 18(4): 267-275.

Guinon, M. 1993. Promoting gene conservation through seed and plant procurement. In: Landis, T.D., ed. Proceedings, Western Forest Nursery Association. Gen. Tech. Rep. RM-221. Fort Collins, CO: USDA Forest Service, Rocky Mountain Forest and Range Experiment Station: 38-46.

Hallman, R. 1993. Reforestation equipment. Publication No. TE02E11. Missoula, MT: USDA Forest Service, Technology and Development Program. 268 p.

Hines, F.D.; Long, A.J. 1986. First- and second-year survival of containerized Engelmann spruce in relation to initial seedling size. Canadian Journal of Forest Research 16: 668-670.

Hoag, J.C.; Landis, T.D. 2001. Riparian zone restoration: field requirements and nursery opportunities. Native Plants Journal 2(1): 30-35.

Jobidon, R.; Roy, V.; Cyr, G. 2003. Net effect of competing vegetation on selected environmental conditions and performance of four spruce seedling stock sizes after eight years in Quebec (Canada). Annals of Forest Science 60: 691-699.

Landis, T.D. 1999. Seedling stock types for outplanting in Alaska. In: Alden, J., ed. Stocking standards and reforestation methods for Alaska. Misc. Publication 99-8. Fairbanks, AK: University of Alaska Fairbanks, Agricultural and Forestry Experiment Station: 78-84.

Landis, T.D. 2001. The target seedling concept: the first step in growing or ordering native plants. In: Haase, D.L.; Rose, R. eds. Native plant propagation and restoration strategies, proceedings of the conference. Portland, OR: Western Forestry and Conservation Association: 71-79.

Landis, T.D. 2002. The target seedling concept: a tool for better communication between nurseries and their customers. In: Riley, L.E.; Dumroese, R.K.; Landis, T.D., tech. coords. National Proceedings: Forest and Conservation Nursery Associations—2002. Proceedings RMRS-P-28. Ogden, UT: USDA Forest Service, Rocky Mountain Research Station: 12-16.

Landis, T.D.; Dreesen, D.R.; Dumroese, R.K. 2003. Sex and the single *Salix*: considerations for riparian restoration. Native Plants Journal 4(2): 110-117.

Lopushinsky, W.; Max, T.A. 1990. Effect of soil temperature on root and shoot growth and on budburst timing in conifer seedling transplants. New Forests 4(2): 107-124.

Mexal, J.G.; Timmis, R.; Morris, W.G. 1979. Cold-hardiness of containerized loblolly pine seedlings: its effect on field survival and growth. Southern Journal of Applied Forestry 3(1): 15-19.

Millar, C.I.; Stephenson, N.L.; Stephens, S.L. 2007. Climate change and forests of the future: managing in the face of uncertainty. Ecological Applications 17(8): 2145-2151.

Nursery Technology Cooperative. 2005. Rapid response reforestation: comparison of one-year-old stocktypes for fire restoration. NTC Annual Report. Corvallis, OR: Oregon State University, Department of Forest Science: 23-27.

Revel, J.; Lavender, D.P.; Charleson, L. 1990. Summer planting of white spruce and lodgepole pine seedlings. FRDA Report 145. Victoria, BC, Canada: Pacific Forestry Centre. 14 p.

Rose, R.; Haase, D.L. 1995. The target seedling concept: implementing a program. In: Landis, T.D.; Cregg, B., tech. coords. National Proceedings, Forest and Conservation Nursery Associations—1995. Gen. Tech. Rep. PNW-GTR-365. Portland, OR: USDA Forest Service, Pacific Northwest Research Station: 124-130.

Rose, R.; Campbell, S.J.; Landis, T.D. 1990. Target seedling symposium: Proceedings, Western Forest Nursery Associations. Gen. Tech. Rep. RM-200. Fort Collins, CO: USDA Forest Service, Rocky Mountain Forest and Range Experiment Station. 286 p.

Scott, G.L. 2006. Personal communication. Missoula, MT: USDA Forest Service, Regional Office.

South, D.B.; Mexal, J.G. 1984. Growing the "best" seedling for reforestation success. Forestry Department Series 12. Auburn, AL: Auburn University, School of Forestry and Wildlife Sciences. 11 p.

South, D.B.; Mitchell, R.J. 1999. Determining the "optimum" slash pine seedling size for use with four levels of vegetation management on a flatwoods site in Georgia, U.S.A. Canadian Journal of Forest Research 29(7): 1039-1046.

St. Clair, B.; Johnson, R. 2003. The structure of genetic variation and implications for the management of seed and planting stock. In: Riley, L.E.; Dumroese, R.K.; Landis, T.D., tech. coords. National Proceedings: Forest and Conservation Nursery Associations—2003. Proceedings RMRS-P-33. Ogden, UT: USDA Forest Service, Rocky Mountain Research Station: 64-71.

Steinfeld, D.E.; Landis, T.D.; Culley, D. 2002. Outplanting long tubes with the Expandable Stinger: a new treatment for riparian restoration. In: Dumroese, R.K.; Riley, L.E.; Landis, T.D., tech. coords. National Proceedings, Forest and Conservation Nursery Associations—1999, 2000, and 2001. Proceedings RMRS-P-24. Fort Collins, CO: USDA Forest Service, Rocky Mountain Research Station: 273-276.

Sutton, R. 1980. Evaluation of stock after planting. New Zealand Journal of Forestry Science 10(1): 297-299.

Van Eerden, E. 2002. Forest nursery history in western Canada with special emphasis on the province of British Columbia. In: Dumroese, R.K.; Riley, L.E.; Landis, T.D., tech. coords. National proceedings: Forest and Conservation Nursery Associations—1999, 2000, and 2001. Proceedings RMRS-P-24. Fort Collins, CO: USDA Forest Service, Rocky Mountain Research Station: 152-159.

White, B. 2003. Container handling and storage in eastern Canada. In: Riley, L.E.; Dumroese, R.K.; Landis, T.D., tech. coords. National Proceedings: Forest and Conservation Nursery Associations—2003. Proceedings RMRS-P-33. Ogden, UT: USDA Forest Service, Rocky Mountain Research Station: 10-14.

The Container Tree Nursery Manual

Volume Seven

Chapter 2
Assessing Plant Quality

by Gary A. Ritchie, Thomas D. Landis, R. Kasten Dumroese, and Diane L. Haase

Contents

7.2.1 **Introduction** *19*

7.2.2 **Classes of Plant Quality Attributes** *20*

7.2.3 **Morphological Attributes** *21*
7.2.3.1 Introduction *21*
7.2.3.2 Morphological characteristics of container seedlings *21*
 Container volume
 Stem diameter ("caliper")
 Shoot height
 "Rootbound" plugs
 Other morphological indices
7.2.3.3 Effects of container size on outplanting performance *24*
7.2.3.4 Morphological attributes: Summary *25*

7.2.4 **Physiological Attributes** *27*
7.2.4.1 Plant moisture stress (PMS) *27*
 What is PMS?
 Water potential
 Units of water potential
 Diurnal patterns of plant water potential
 Measurement of plant moisture stress
 Interpretation of PMS values
 Is PMS an indicator of plant quality?
 PMS as a snapshot of plant water status
 Plant moisture stress: Summary
7.2.4.2 Cold hardiness *33*
 Concepts behind the test
 What happens when plant tissues freeze?
 Cold hardiness mechanism
 Stages in cold hardening
 Hardiness variation in plant tissues, species, and ecotypes
 Cold hardiness testing methods
 Whole plant freezing test
 Freeze-induced electrolyte leakage test
 Differential thermal analysis
 Cold hardiness testing through gene expression
 Applications of cold hardiness testing
 Cold hardiness: Summary
7.2.4.3 Root electrolyte leakage *40*
 Theory
 The biological significance of REL
 Measurement procedure
 Applications of REL in nurseries
 REL as a predictor of outplanting performance
 Limitations of REL
 Root electrolyte leakage: Summary

7.2.4.4 Chlorophyll fluorescence *44*
 What is chlorophyll fluorescence?
 Photosynthesis and chlorophyll fluorescence
 Measuring chlorophyll fluorescence
 Normal values of CF parameters in plants
 Use of CF in plant-quality assessment
 Chlorophyll fluorescence: Summary
7.2.4.5 Mineral nutrient content *48*
7.2.4.6 Carbohydrate reserves *50*

7.2.5 **Performance Attributes** *52*
7.2.5.1 Bud dormancy *52*
 The concept of dormancy
 Defining dormancy
 The dormancy cycle
 The chilling requirement
 Measuring dormancy
 Calculating the dormancy release index
 Measuring mitotic index
 Bud size and development
 Dormancy: Summary
7.2.5.2 Stress resistance *58*
 The concept of stress resistance
 Measuring stress resistance
 Using cold hardiness tests to estimate overall stress resistance
 Using chilling hours to predict stress resistance
 Adjusting for the added effect of refrigerated storage
 Application to other species and regions
 Stress resistance: Summary
7.2.5.3 Root growth potential *62*
 RGP test procedure
 RGP as a predictor of field performance
 Why RGP often works
 Root growth potential: Summary

7.2.6 **Correlating Combinations of Plant Quality Tests to Predict Outplanting Performance** *68*

7.2.7 **Limitations of Plant Quality Tests** *69*
7.2.7.1 Timing *68*
7.2.7.2 Sampling *68*
7.2.7.3 Unreasonable expectations *70*

7.2.8 **Commercial Plant Quality Testing Laboratories** *71*

7.2.9 **Summary and Conclusions** *72*

7.2.10 **Literature Cited** *74*

7.2.11 **Appendix** *81*

7.2.1 Introduction

In his prophetic work "Planting the Southern Pines," Wakeley (1954) foresaw what we now hold as axiomatic—restoration, including forestation, can never be entirely successful until nurseries are able to produce crops of "high-quality" plants consistently and reliably. But how to distinguish a high-quality plant from a low-quality one was not always obvious, so the concept of plant quality remained obscure for many years. Wakeley also recognized that "morphological grades" often fell short in their ability to predict performance, and he hypothesized that "physiological grades" may be a better criterion of viability (Wakeley 1949). What exactly constituted a physiological grade, however, and how to measure it, eluded Wakeley and his contemporaries.

During the past 30 years, worldwide, nursery researchers and managers convened numerous workshops and symposia and published many reports on the subject of plant quality and how to measure it (for example, Colombo 2005; Duryea 1985; Haase 2008). This work generated a variety of quality tests; although many are ingenious, most failed to meet expectations. A few, however, stood the test of time and remain in operational use. In this chapter, we discuss the most practical ways of measuring plant quality and how these methods can be used in container nurseries.

7.2.2 Classes of Plant Quality Attributes

Forestry researchers have labored to identify quantifiable traits that could be used as indicators of plant quality and, better yet, predictors of performance after outplanting. Although an impressive list of such attributes has been assembled (for example, Grossnickle 2000), relatively few are used operationally in nurseries or on the outplanting site. In our view, plant quality can be divided into three broad classes:

Morphological attributes—These traits can be readily seen and easily measured, such as stem height, stem (root collar) diameter, root volume, and root and shoot dry weight. During the harvesting-to-outplanting process, these traits do not change appreciably.

Physiological attributes—These traits are not readily visible and need to be measured with instruments or through laboratory procedures. In contrast to morphological characteristics, physiological attributes change often and sometimes dramatically during the harvesting-to-outplanting process. Therefore, any measurement of physiological quality is a "snapshot," relevant for only a brief point in time. Some common physiological attributes include cold hardiness and bud dormancy.

Performance attributes—These traits can be assessed only by subjecting plants to certain predefined testing protocols and observing how they perform. Performance tests have great value because they assess and integrate a wide spectrum of morphological and physiological traits at once. Unfortunately, performance tests are laborious, time consuming, and therefore expensive. Nevertheless, because of their intuitive appeal, performance tests have found wide use in plant quality assessment. One of the oldest and still most commonly used performance tests is the root growth potential test.

7.2.3 Morphological Attributes

7.2.3.1 Introduction

Most nursery stock produced in the United States, Canada, and Europe during the 1970s was bareroot, and so most seedling morphology literature focuses on bareroot stocktypes (Frampton and others 2002; Ritchie and others 1997). The effects of morphology on performance of bareroot stock have been summarized in the literature (Mexal and Landis 1990; Thompson 1985; Wilson and Jacobs 2006); height, stem diameter, root system "quality" (volume or mass), and the ratio of the mass of the shoot to that of the root system are typically the best predictors of outplanting performance. Survival is best forecast by stem diameter, while shoot growth tends to be more related to initial seedling height. With bareroot stock, when stem diameter increases above about 5 mm (0.2 in.), other morphological indicators become less important (Mexal and Landis 1990). In addition, bareroot seedlings with larger root volumes at the time of outplanting have greater subsequent growth and survival than those with smaller root volumes (Rose and others 1997).

7.2.3.2 Morphological characteristics of container seedlings

Let's discuss, in order of importance, the major morphological factors that describe container stock quality.

Container volume. The most important morphological factor affecting plant quality in container nurseries is container size or volume. Container volume controls the amount of roots that a plant can produce, which in turn, determines how large a shoot can be produced in a given amount of time. In addition, the size of the container "plug" limits the moisture and mineral nutrient reserves that will be taken to the outplanting site. Compared with bareroot stock that has extremely variable root systems, it is easy to characterize the volume and depth of container root plugs; most container nursery stock is described by container volume. For example, in the Northwestern United States, a "Styro 20" refers to a plant that has been produced in a Styrofoam™ block container with cells that are 340 cm^3 (20 in^3) in volume.

Container volume is the most important factor controlling root egress after outplanting (fig. 7.2.1A). As container volume increases, the amount of exterior surface area of the root plug also increases (fig. 7.2.1B), which means that plugs of larger containers have more surface contact with the surrounding soil.

Among different container sizes, volume and growing density have the most significant effect on plant morphology (table 7.2.1). In studies with interior spruce (*Picea glauca* x *engelmanii* complex) (Grossnickle 2000); Douglas-fir, western hemlock (*Tsuga heterophylla*), and Sitka spruce (*Picea sitchensis*) (Arnott and Beddows 1982); black spruce (*Picea mariana*) (Jobidon and others 1998); and cherrybark oak (*Quercus pagoda*) (Howell and Harrington 2004), every morphological trait measured increased in value as container volume increased. In every case, container stock with larger root plugs grew larger after outplanting.

Because block containers have fixed cell spacing, it is more difficult to study the effects of changing plant density at the same cell volume. In contrast, the Ray Leach Conetainer® system allows cell spacing to be changed, allowing a few good research trials to be done. Douglas-fir seedlings grown at densities ranging from 270 to 1,080 plants/m^2 (25 to 100/ft^2) showed that shoot height increased with increasing density because of the competition for light in response to crowding (fig. 7.2.2). Stem diameter decreased, however, which shows that quality can be lessened by growing plants too closely together (Timmis and Tanaka 1976).

Within containers of the same size, stem diameter and shoot height have proved to be the most important morphological traits affecting quality and, therefore, are the two factors most often used in grading specifications (fig. 7.2.3A). More discussion on measuring height and stem diameter is provided in Volume One, Section 1.5.4.2.

Stem diameter ("caliper"). Stem diameter is typically measured, using a small caliper, at the root collar where the stem meets the root system. Root-collar diameter, or stem diameter, is always reported in millimeters. Numerous studies show that stem diameter is the best predictor of outplanting performance and, therefore, plant quality. When Engelmann spruce container seedlings with a range of stem diameters were outplanted on a high elevation site in Utah, survival after two growing seasons was strongly correlated with initial stem diameter (fig. 7.2.3B). This information

Figure 7.2.1—*Root growth out of the plug and into the surrounding soil ("egress") is critical to plant survival and growth after outplanting (A). Container volume is important not only because it determines the amount of roots that a container plant has, but also because the surface area of the plug is in contact with the surrounding soil (B) (A, modified from Grossnickle 2000).*

Figure 7.2.2—*When plants are grown in the same volume container but at different densities, shoot height increases with closer spacing whereas stem diameter decreases (modified from Timmis and Tanaka 1976).*

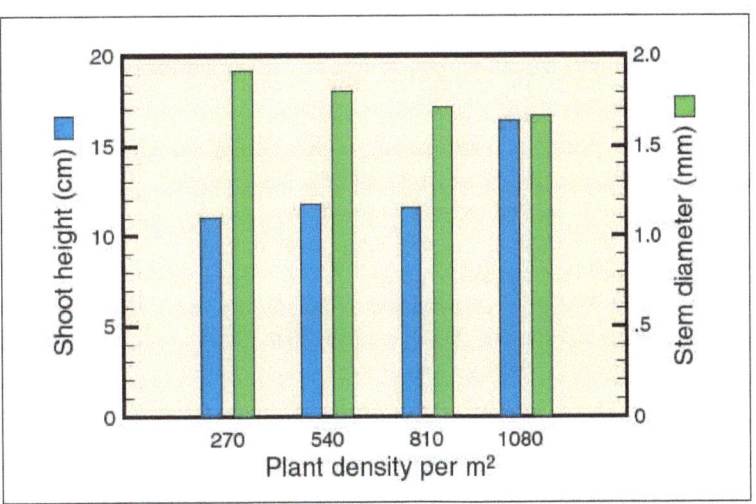

Table 7.2.1—*Effect of container volume on seedling morphology of 2-year-old interior spruce (Picea glauca x Picea engelmannii)**

Seedling morphological attributes	Styroblock™ cell volumes		
	105 cm³ (6.6 in³)	170 cm³ (10 in³)	340 cm³ (20 in³)
Shoot height–cm (in)	24.2 (9.5)	29.7 (11.7)	33.3 (13.1)
Root-collar diameter–mm	4.4	5.0	6.8
Shoot dry weight–g (oz)	2.8 (0.10)	4.5 (0.16)	6.4 (0.23)
Root dry weight–g (oz)	1.1 (0.04)	1.4 (0.05)	2.1 (0.07)
Number of branches	18	24	33
Number of buds	50	67	86

* Source: Grossnickle (2000).

was used to develop grading standards; in this case, seedlings with stem diameters ≥ 2.5 mm were shippable, whereas smaller ones were not (Hines and Long 1986). Of course, this relationship varies with conditions on the outplanting site so standards must be developed for each species and for different outplanting conditions.

Shoot height. Height is the distance from the root collar to the tip of the terminal bud or shoot. It is usually reported in centimeters or millimeters, but in the United States it is often reported in inches. This results in the peculiar situation in which plants are characterized using both English and metric measuring systems; for example, a plant shoot that is 12 inches high with a 5 mm caliper. Height is correlated with the number of needles on the stem and, therefore, is a good estimate of photosynthetic capacity and transpirational area.

"Rootbound" plugs. The fact that excessive root growth becomes a quality issue in container plants has been known for decades but, until recently, no morphological index or rating system had been developed. Rootbound nursery stock can be defined as plants that have grown too large for their container, resulting in severe matting and tangling of the root system (fig. 7.2.4A). From a quality standpoint, this condition reduces plant survival or growth after outplanting (South and Mitchell 2006). Several studies have related rootbinding to the length of time that the plant has been in the container. Usually, the larger the container, the longer it takes for the plant to become rootbound. But time alone is not really useful, because root growth is also affected by cultural conditions at the nursery. A species growing rapidly in one nursery will become rootbound faster than the same species growing more slowly in another nursery. Similarly, a species in a large container given large amounts of fertilizer may become rootbound as fast as the same species in a smaller container given smaller amounts.

When plants are grown in the same volume container, outplanting survival has been shown to decrease after an optimum root-collar diameter is exceeded (fig. 7.2.4B). South and Mitchell (2006) propose a "root-bound index" based on root-collar diameter divided by container diameter or volume that must be calculated for each container type. From an operational standpoint, however, establishing a maximum stem diameter along with a visual assessment of root binding might be the most practical culling system.

Other morphological indices. Several other morphological criteria, such as biomass, shoot-to-root ratio, sturdiness, and appearance, have been used to describe plant quality. **Biomass** can be determined using volume or dry weight methods. Shoots and roots are usually measured separately. Dry weight is determined by cleaning, oven drying,

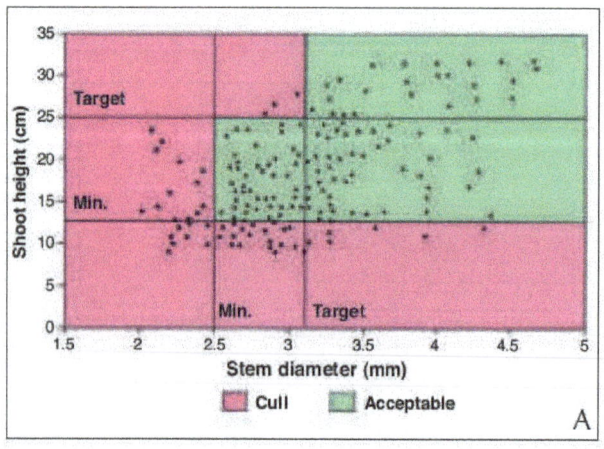
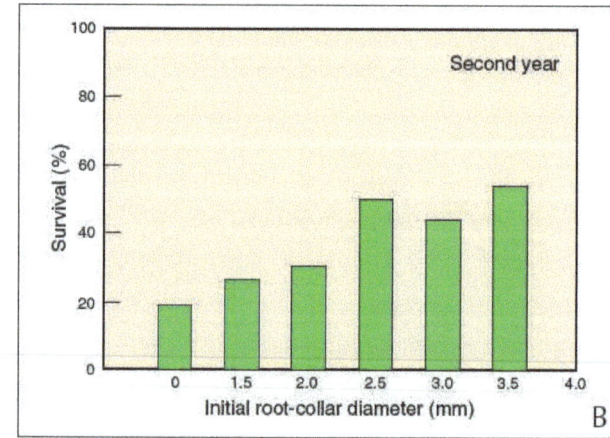

Figure 7.2.3—*Shoot height and stem diameter are the most common grading criteria in container nurseries (A), but stem diameter has proven to be the best single morphological indicator of seedling quality. When Engelmann spruce (Picea engelmannii) container stock were outplanted, plants with stem diameters larger than 2.5 mm outperformed smaller ones after the second year (B) (modified from Hines and Long 1986).*

and weighing plants. Volumes are determined using water displacement (Burdett 1979; Harrington and others 1994). **Shoot-to-root ratio** (shoot:root) is the ratio of the dry mass or volume of the shoot to the dry mass or volume of the root system and provides an indicator of the "balance" of the plant. When the shoot:root is "1" the size of the root mass equals the size of the shoot mass. More often, however, the ratio is greater than 1 because the shoot often outweighs the root system. Shoot-to-root ratios less than 2.5 are usually deemed most desirable. A **sturdiness ratio** is calculated by dividing shoot height (cm) by diameter (mm). It attempts to capture the idea of "sturdiness" (low value) in contrast to "spindliness" (high value). This ratio has found particular use in container stock, which can become tall and thin when grown at high densities and/or under lower lighting conditions. **Color, form, and damage** should also be accounted for when evaluating morphological quality. Foliar color is a general indicator of plant quality and can vary by species and time of season. Yellow, brown, or pale-green foliage indicates lower vigor and/or chlorophyll content than dark green foliage. The foliage of some species turns purple during winter dormancy, but this is not considered diagnostic (see Section 7.2.5.1). Multiple shoots, stem sweep, root deformity, physical damage, and any other noticeable characteristics that can affect plant performance are also important factors to note when assessing morphological quality. A single, but comprehensive study with Italian stone pine (*Pinus pinea*) container seedlings measured various morphological characteristics. The best single indicator of plant quality was the **ratio of container depth to stem diameter**, and target plants had a container depth-to-stem diameter ratio of 4 (Dominguez-Lerena and others 2006).

7.2.3.3 Effects of container size on outplanting performance

The main objective of measuring plant morphological traits is to predict performance after outplanting—specifically survival and growth.

So what traits, or combinations of traits, have the greatest positive effect on plant performance? The conventional wisdom is that bigger is generally better than smaller. All other factors being equal, big plants with proportionately larger stem diameters and root systems normally exhibit higher survival and greater growth than smaller plants or

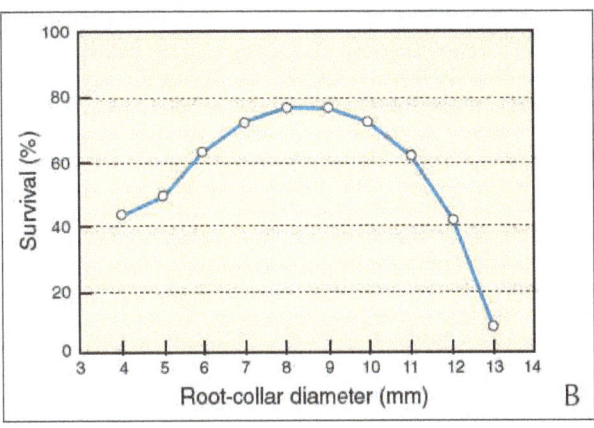

Figure 7.2.4—*Container plants that have grown too long in the same container become "rootbound" which greatly reduces their quality (A). For a given species and container size, an optimum stem diameter exists that can be used for grading-out rootbound plants; this figure was developed for longleaf pine* (Pinus palustris) *(B) (B, modified from South and Mitchell 2006).*

those with poorly developed roots. In general, outplanting survival is more related to stem diameter, whereas shoot growth after outplanting depends more on initial plant height (Arnott and Beddows 1982).

As discussed in Chapter 7.1, survival and growth also depend strongly on environmental conditions of the outplanting site. After reviewing the literature on container size and performance, Grossnickle (2005) concluded that "large" seedlings performed better than "small" seedlings on moist sites where vegetative competition was severe. Conversely, smaller seedlings fared better on sites prone to water stress. On sites with heavy vegetative competition, the ability to access and process sunlight strongly determines survival and growth. Hence taller, branchier seedlings with a large photosynthetic area have an advantage over smaller seedlings that tend to become shaded by competing vegetation. For example, large white spruce (*Picea glauca*) seedlings outplanted in a boreal British Columbia forest were better equipped for competition than were smaller seedlings (McMinn 1982). Similarly, tall container seedlings of Douglas-fir, western hemlock, and Sitka spruce exhibited greater height growth after outplanting on a coastal British Columbia site than shorter seedlings (Arnott and Beddows 1982). In a study in Quebec, large spruce seedlings grew better than did smaller ones on mesic sites with heavy plant competition (fig. 7.2.5). Larger stock with thick stems also performs better on sites with animal predation and heavy snow, as shown with Engelmann spruce seedlings (Hines and Long 1986).

Contrast this with an outplanting site where hot and dry conditions cause high evapotranspirational demand. Here, the advantage lies with plants that have a relatively small transpirational surface area relative to a large, absorptive root system. Under these conditions, nursery plants with a large shoot and small root system (high shoot:root) are at a disadvantage because they transpire faster than they can absorb water from the soil. For these high-stress sites, specifying larger volume containers at lower growing densities (wider cell spacing) will produce plants that have a short shoot and thick stem diameter (Grossnickle 2005).

Miniplug transplants are a stocktype that results in large plants in a relatively short time (Landis 2007). Growers sow miniplugs (approximately 16 cm^3 [1 in^3] cavity volume) in a greenhouse during mid-winter and then transplant them a few months later to larger, wider spaced containers that are moved to outdoor growing areas or bareroot nursery beds. These "plug + plug" transplants have proved to be popular stocktypes for hot and dry outplanting sites (fig. 7.2.6).

Although much less research has been done on broadleaved (hardwood) species, the review by Wilson and Jacobs (2006) notes that, as with conifers, height and stem diameter are the most frequently used grading criteria for hardwoods, with stem diameter usually providing the most consistent prediction of field performance.

7.2.3.4 Morphological attributes: Summary

Shoot height and stem diameter are the most frequently measured morphological traits and the most common grading criteria. Morphological attributes are easily assessed and do not change appreciably during the harvesting-to-outplanting process. Nearly all morphological traits reflect container volume and/or growing density; large container volumes and low growing density promote development of large stock.

Effects of morphology on performance of container stock mirror those for bareroot stock:

- Initial stem diameter tends to be correlated with survival.
- Initial height tends to be correlated with shoot growth.
- Morphological traits can interact. For example, stem diameter may influence survival in plants with poor root systems but not in those with good root systems.
- Larger stock generally performs better than smaller-stock, but this depends on conditions on the outplanting site. Tall stock with thick, stiff stems and a large photosynthetic surface is best for sites with plant competition, animal predation, or heavy snow loads. Short stock with thick, stiff stems and extensive root systems is best for droughty sites.

As discussed earlier, physiological traits of nursery plants differ significantly from morphological characteristics in that they are not readily seen, they change often and sometimes dramatically throughout the harvest-to-outplanting process, and they must be measured with laboratory equipment.

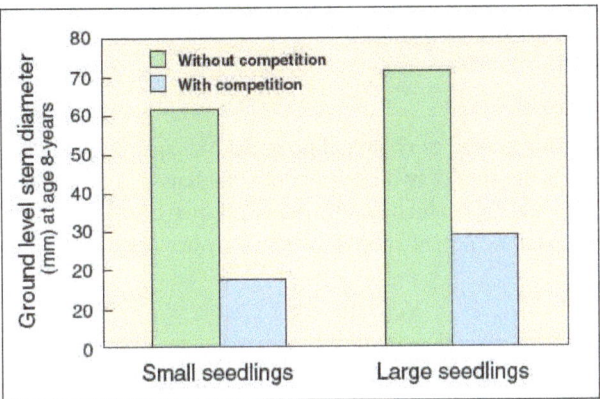

Figure 7.2.5—*Large black spruce and white spruce container stock outperformed smaller plants when measured 8 years after outplanting in southeastern Quebec (modified from Thiffault 2004).*

Most physiologically based quality tests measure only one plant function, such as cold tolerance, water status, or photosynthetic efficiency. It is helpful to think of plant quality in layers: morphological characteristics are the base layer, whereas physiological traits are the second layer. A batch of plants may have ideal shoot height and stem diameter, but these morphological traits alone are insufficient to guarantee high quality. Physiological tests are needed to provide a more comprehensive picture.

In the next section, we discuss four tests of physiological quality: plant moisture stress, cold hardiness, root electrolyte leakage, and chlorophyll fluorescence.

Figure 7.2.6—*For hot and dry outplanting sites, these "Q-plug + one" Jeffrey pine (Pinus jeffreyi) have the ideal morphology — short stocky shoots (A) with large stem diameters and root mass (B).*

7.2.4 Physiological Attributes

7.2.4.1 Plant moisture stress (PMS)

Plant moisture stress, or PMS, is one of the oldest and most commonly used tests to measure quality. Its popularity rests on its simplicity and robustness, and the fact that PMS equipment is relatively inexpensive, intuitive, and portable. Although PMS measurements are easily made, their interpretation can be more difficult.

What is PMS? Without a steady supply of good quality water, plants cease growing and ultimately die. The amount of water needed to meet the basic metabolic needs of a plant is quite low. During photosynthesis, atmospheric carbon dioxide (CO_2) diffuses into leaves through stomata and, once inside the leaf, this CO_2 is converted to sugars. Photosynthesis is, however, a very "leaky" process because, while CO_2 is diffusing into the leaves, water is diffusing out—this loss of water is called transpiration. Plants can reduce transpiration by closing stomata, but this impedes photosynthesis. So, in order to grow, plants must transpire vast amounts of water.

Transpiration generates a tension (or "stress") that due to water's high cohesion, is transmitted through vascular tissue from the leaf down through the stem and into the roots. During daylight, when stomata are open, transpiration typically exceeds the plant's ability to extract water from the soil. Therefore, during the day, plants are always under some degree of water stress. This stress is perfectly normal and not injurious unless it reaches high levels for a prolonged period of time.

In very simple terms, plant moisture stress can be modeled as:

$$PMS = A - T + S$$

where A is the absorption of water from the soil, T is transpirational loss, and S is storage of water in the plant's stem and roots, which is negligible in seedlings but important in large trees. During daylight, T almost always exceeds A.

Water potential. A more precise way to model the state of water in plants is the thermodynamic approach, which is based on water potential and represented by the Greek letter psi (ψ). The total water potential (ψ_W) is a measure of the free energy or chemical potential of water. In plants, ψ_W is the sum of two component potentials: the pressure potential (ψ_P), which can be either positive or negative, and the osmotic potential (ψ_O), which is always negative:

$$\psi_W = \psi_P + \psi_O$$

Potentials are expressed in units of pressure and, although MegaPascals (MPa) are the official SI units, bars are most commonly used by nursery and reforestation personnel. By definition, the ψ_W of pure water at standard temperature and pressure is 0 bars, or 0 MPa. ψ_P and ψ_O are continually changing as transpiration and osmosis cause water to move across membranes, in and out of cells, and up the transpiration stream.

The components of water potential have different properties depending on where the water is located within the plant tissues. Water is contained within cell membranes as part of the symplast and outside cell membranes as part of the apoplast. In the apoplast, water is nearly always under hydrostatic tension from transpirational pull, so pressure potential (ψ_P) is always negative (table 7.2.2). In the symplast, however, ψ_P is normally positive owing to the inward turgor pressure that cell membranes and walls exert on cell contents. The exception would be for a cell that has lost all turgor (wilted), in which case $\psi_P = 0$. This is often called the "zero turgor point," which is discussed below. The osmotic component (ψ_O) is normally near 0 in the apoplast whereas, in the symplast, ψ_O is always negative owing to effects of dissolved solutes (ions) in the cells (table 7.2.2). These component potentials are continually changing as water moves across cell membranes due to osmosis or up through the plant due to transpiration. Because ψ_W is the sum of these two components, it is almost always negative and the plant is almost always under some level of water deficit, or stress.

The interplay of these component potentials in the symplast can be visualized with a Höfler diagram (fig. 7.2.7). The X-axis is the water content of the cell expressed as a percentage of full turgor. The Y-axis gives the component potentials. At full hydration (A in figure 7.2.7), plants are turgid and the positive turgor pressure of cell walls (ψ_P) balances the negative osmotic potential (ψ_O) of cell contents. At this point, $\psi_W = 0$ MPa. As cells lose water, ψ_P falls and the concentration of solutes in cells increases.

This drives ψ_O down, so ψ_W also falls. When ψ_P reaches 0 MPa (B in figure 7.2.7), cells collapse and plants wilt. The value of ψ_W at which this occurs is known as the "zero turgor point" or, as it is more commonly known, the "permanent wilting point" (C in figure 7.2.7).

Units of water potential. Thermodynamic water potential terminology (Slatyer 1967) has sometimes been troublesome for growers because negative values are hard to visualize and tricky to manipulate algebraically. For this reason, water potential is often expressed as a positive value and is called "Plant Moisture Stress" (PMS). These values can be easily converted because –1.0 MPa equals 10 bars. This relationship and some examples are shown in table 7.2.3. For example, a PMS value of 10 bars indicates a "moderate" level of stress and is equivalent to ψ_W of –1.0 MPa. From a theoretical standpoint, however, thermodynamic terminology is useful because it is consistent through the soil-plant-atmosphere continuum (fig. 7.2.8).

Diurnal patterns of plant water potential. As already mentioned, ψ_W is dynamic and this affects its usefulness as an index of plant quality. Consider, for example, a container plant whose growing medium is at field capacity with water. During daylight, while stomata are open, low humidity (high vapor pressure deficit) draws moisture from the leaves. This creates an imbalance between transpiration and water absorption, resulting in the development of PMS at midday (ψ_W decreases). During nighttime, stomata tend to close, relative humidity rises to nearly 100 percent, and transpiration ceases. The negative ψ_W in the plant pulls water from the soil or growing medium, thereby relieving the stress. By early the next morning, pre-dawn ψ_W reaches a dynamic equilibrium with soil moisture potential ($\psi_W = \psi_{soil}$).

If no water is added to the container, the growing medium dries out, and predawn and midday plant moisture stress increase daily as ψ_{soil} decreases. After a few days, the plant will close its stomata during midday to retard transpiration. This can be seen occurring in days 4 and 5 in figure 7.2.9, and results in a moderating of the midday PMS. ψ_{soil} will eventually become so negative that the plant will be unable to equilibrate during the night. Throughout this time, the midday stress will continue to increase. When irrigated, the system will return to the ini-

Table 7.2.2—*Properties of component water potentials in the symplast and apoplast*

Component potential	Apoplast (outside cells)	Symplast (inside cells)
Pressure potential (ψ_P)	Always negative	Generally positive Zero when wilted
Osmotic potential (ψ_O)	Generally slightly negative	Always negative
Water potential (ψ_W)	Always negative	Variable

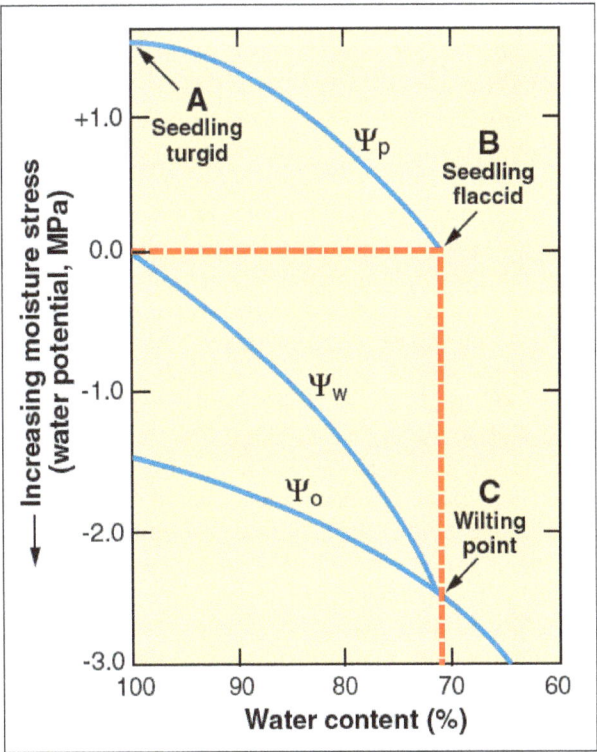

Figure 7.2.7—*The interrelationships between plant water potential (ψ_W) and its components, osmotic potential (ψ_O) and pressure potential (ψ_P), change over the range of plant water contents from turgidity (A) to the permanent wilting point (PWP) (C) (modified from Ritchie 1984b).*

tial state shown in day 1 unless the plant has experienced irreversible damage from the high PMS.

Note that the ability to track moisture stress levels of both soil and plant in figure 7.2.9 shows the advantage of using water potential units rather than PMS, which reflects only plant stress.

Measurement of plant moisture stress. Over the years, as plant physiologists labored to understand the dynamics of plant water relations, many attempts were made to develop methods of measuring ψ_w (Lopushinsky 1990). As far as nursery work goes, the most significant development was invention of the "Scholander Pressure Chamber" (Scholander and others 1965), based on an earlier glass pressure chamber devised by Dixon (1914). Waring and Cleary (1967) modified the chamber for trees and seedlings and outlined basic measurement procedures.

The modern pressure chamber consists of a metal pressure vessel connected to a nitrogen gas source through a pressure regulator. To measure plant moisture stress, the stem is cut and inserted through a rubber or compression gasket. A new model pressure chamber from the PMS Instrument Company comes equipped with a "rubber gland" instead of a gasket, which greatly improves the speed and accuracy of measurements. This is then sealed into a hole in the chamber lid with the foliage inside the chamber and the cut stem protruding (fig. 7.2.10). Nitrogen gas is slowly bled into the chamber while the cut stem is closely observed. When a droplet of water appears at the end of the stem, the chamber pressure is noted. The gas pressure required to force water to the surface is equal to the moisture stress of the plant. For a detailed theoretical description and procedural guide see Ritchie and Hinckley (1975).

The pressure chamber is the standard technique used for measuring PMS in forest nurseries, on outplanting sites, and in plant research facilities. For example, the Forest Service J.H. Stone Nursery in Central Point, Oregon, uses pressure chambers to measure PMS for scheduling bareroot seedling irrigation and to detect dangerous PMS levels during lifting and packing operations (J.H. Stone Nursery 1996).

Table 7.2.3—*Comparison of units and terms for plant water potential and plant moisture stress (modified from Landis and others 1989)*

Plant water potential (MPa)	Plant moisture stress (bars)	Relative moisture stress rating	Relative moisture condition
0.0	0.0	Very low	Wet
0.5	5.0	Low	
1.0	10.0	Moderate	
1.5	15.0	High	
2.0	20.0	High	
2.5	25.0	Very high	Dry

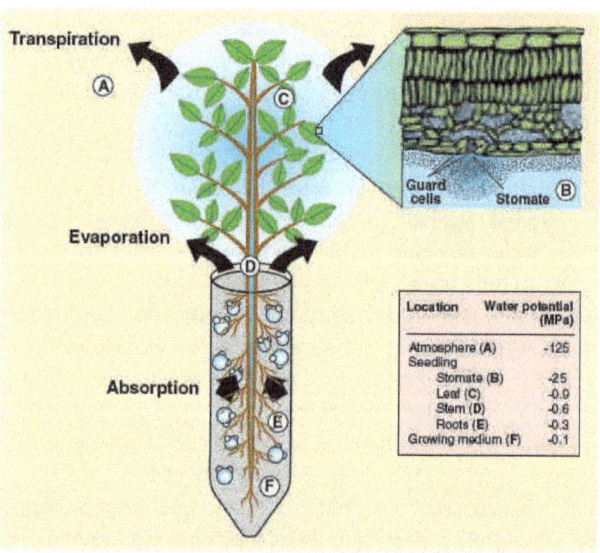

Figure 7.2.8—*Water is pulled along a gradient of water potential that is driven by evapotranspiration, from higher (less negative) levels in the growing medium through the plant to lower (more negative) levels in the surrounding air (modified from McDonald and Running 1979).*

Pressure chambers and supplies are available from the following companies:

PMS Instrument Company
1725 Geary Street SE
Albany, OR 97322 USA
Tel: 541-704-2299
Fax: 541-704-2388
E-mail: info@pmsinstrument.com
Web site: http://pmsinstrument.com/

or

Soil Moisture Equipment Corporation
Santa Barbara, CA
Tel: 805-964-3525 ext. 248
E-mail: alle@soilmoisture.com
Web site: http://www.soilmoisture.com/

Interpretation of PMS values. PMS measurements are used extensively in plant physiology and ecological research because they are robust, easy to obtain, and their relationship to plant physiology is easy to demonstrate. For example, when container white spruce was subjected to extended moisture stress, stomata closed and photosynthesis ceased abruptly at –2 MPa (20 bars) (fig. 7.2.11). Unless this stress is relieved, plant growth will most certainly be restricted and death may occur.

The relationship between PMS readings and plant quality, unfortunately, is not always as straightforward as one might hope. This is partly because PMS, as an estimate of ψ_w, integrates several variables into one reading and, therefore, much information is lost. In addition, because the components of water potential change seasonally, a given value of PMS might have a different interpretation if taken in spring as opposed to winter. For example, figure 7.2.12 shows how the "zero turgor point" changes seasonally in roots and stems of Douglas-fir seedlings (Ritchie and Shula 1984). Looking at stem values, a PMS reading of –2.5 MPa (25 bars) would be a potentially lethal value if taken in April, because it would be near the zero turgor point. But the same value, if measured in January, would be of little concern. On the other hand, root systems with PMS near –2.0 MPa (20 bars) would be suspect most of the year.

As illustrated in figure 7.2.9, PMS can vary sharply throughout the day and from day to day. Daytime PMS

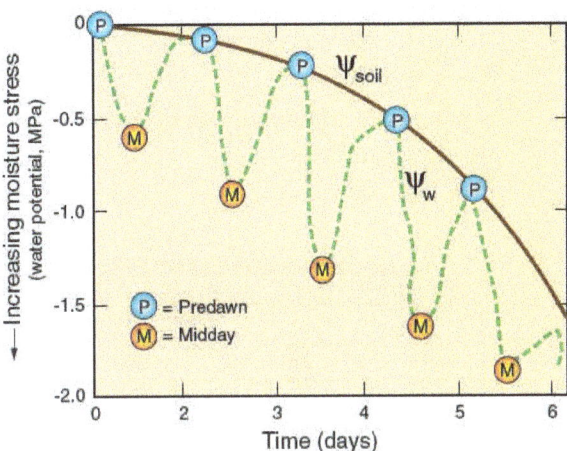

Figure 7.2.9—*For a plant growing in a nonirrigated container, the plant water potential (ψ_w) gradually decreases as the growing medium (ψ_{soil}) dries (modified from Slatyer 1967).*

values can fluctuate widely on days with intermittent sunshine and wind, providing only brief "snapshots" of PMS that have little diagnostic value. Probably the most useful PMS value is what is known as "predawn PMS." This is the PMS that occurs just before sunrise when ψ_w is in dynamic equilibrium with ψ_{soil} (fig. 7.2.9) and provides an estimate of the minimum stress the plant might experience that day. If this minimum value is high, it may be cause for concern. With the above caveats in mind, we present some suggested guidelines for interpretation of predawn PMS measurements as they relate to plant growth and cultural implications (table 7.2.4).

Is PMS an indicator of plant quality? As pointed out by Lopushinsky (1990), the commonly used plant quality indicators (root growth potential, cold hardiness, stress resistance, and dormancy intensity) are not correlated with PMS. Therefore, PMS cannot be used as a proxy indicator of any of these. So, can PMS alone be a useful indicator of quality?

In our opinion, PMS reflects quality only when stress is moderately high and sustained for several days. For example, nursery stock with *predawn* PMS values in the range of –1.5 to –2.5 MPa (15 to 25 bar) range is under severe stress (table 7.2.4), especially if these readings persist after irrigation. We should also point out that dead plants can exhibit very low PMS values because dead roots retain the ability to absorb water. So, low PMS values are not necessarily indicators of healthy stock.

PMS is also used operationally to monitor plant condition during the harvest-to-outplanting process. For example, stock that has a plant water potential (PMS) value of, say, –1.0 MPa (10 bars) coming out of refrigerated storage would certainly be cause for concern. Likewise, nursery stock should have low PMS values immediately before outplanting, when high values indicate overheating or exposure to sun or wind.

You may have noted that all research has been done with conifers. Use of PMS as a performance predictor for deciduous hardwoods also shows some promise, although Wilson and Jacobs (2006) point out that much work is needed to define critical PMS values for a given species.

PMS as a snapshot of plant water status. The fact that PMS is not always a good predictor of plant quality should not be interpreted to mean that monitoring PMS is a waste of time. Pressure chambers should be used to check plant moisture status at several times during nursery tenure. Using predawn PMS readings to fine-tune nursery irrigation practices is a good idea, because pressure chamber measurements show the actual water status of a plant at a given time.

PMS measurements during harvesting can alert nursery managers to dangerously dry conditions or excessive plant exposure (MacDonald and Running 1979). PMS can also be used to check the moisture status of stock immediately before outplanting. For example, a very strong relationship was found between PMS readings taken immediately before outplanting of radiata pine (*Pinus radiata*) seedlings and root growth potential (Mena-Petite and others 2001) (fig. 7.2.13).

Plant moisture stress: Summary. Plants normally lose water more rapidly through transpiration than they absorb from the soil, so they are almost always under some level of water stress, commonly known as plant moisture stress (PMS). PMS is linearly correlated to, but differs in sign from, plant water potential (ψ_w). PMS shows strong diurnal variations as transpiration rates adjust in response to changes in temperature, vapor pressure deficit, and stomatal aperture. The most useful value of PMS is that which

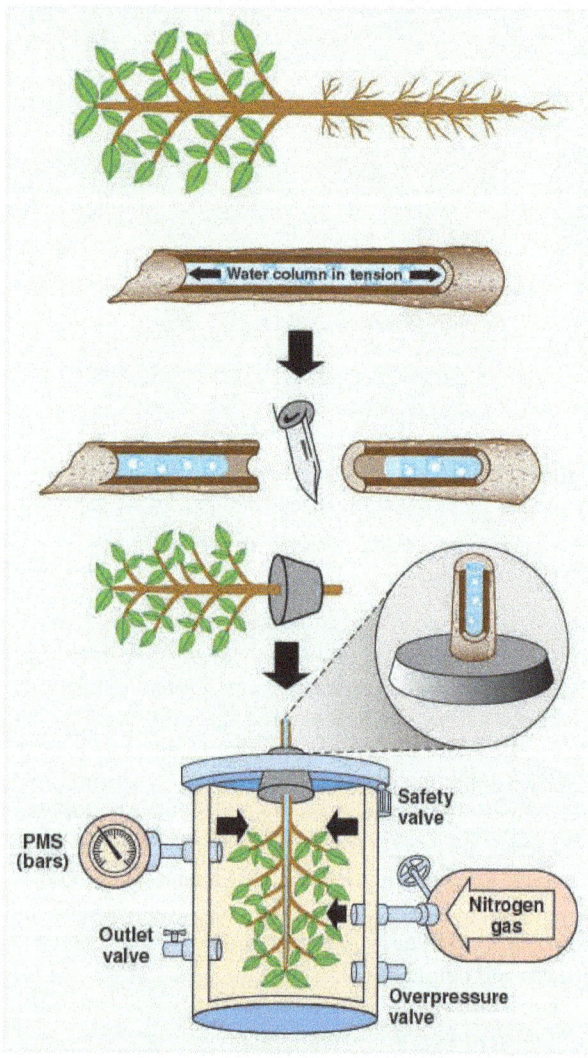

Figure 7.2.10—*How to measure plant moisture stress (PMS) with a pressure chamber. A plant stem is severed and the cut end forced through a hole in the center of a rubber gland, which is then inserted into the lid of the chamber. Nitrogen gas is slowly introduced into the chamber until a drop of water is forced to the surface of the cut stem. The gauge pressure at which this occurs is equal and opposite to the forces holding the water in the stem and is known as PMS.*

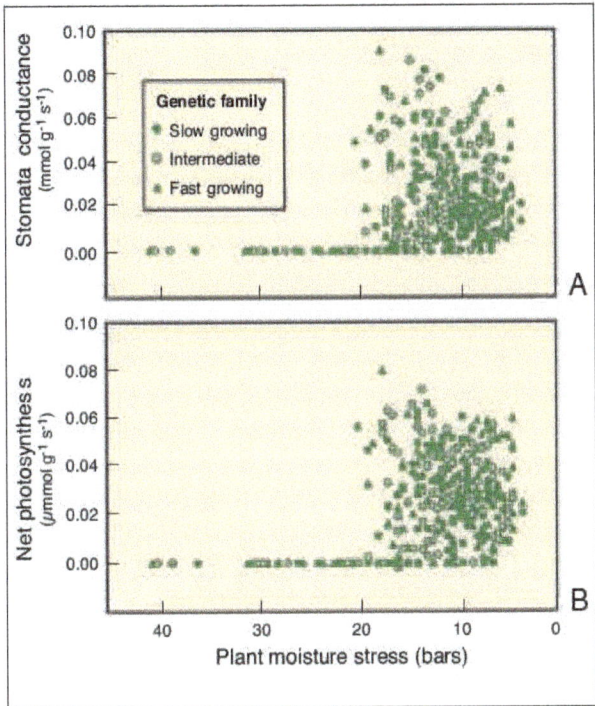

Figure 7.2.11—*Plant moisture stress can give an instantaneous indication of nursery stock water status. When different families of white spruce seedlings were placed under increasing water stress, the stomata closed (A) and all photosynthesis ceased at –2 MPa (20 bars) (B, modified from Bigras 2005).*

occurs just before dawn (pre-dawn PMS), when ψ_w is in near equilibrium with ψ_{soil}. The Scholander pressure chamber, introduced in the mid 1960s, remains the most robust and useful method for measuring PMS. In this test, a stem is severed from a plant, sealed in a pressure chamber, and gas under pressure is introduced into the chamber until a water drop forms at the cut surface. The pressure at which this occurs is equal and opposite to the forces holding the water in the stem and provides an estimate of PMS. Although there are strong seasonal variations in critical PMS (plant water potential) values, readings in the range of –0.5 to –1.5 MPa (5 to 15 bars) are normal whereas those below –1.5 MPa (above 15 bars) can be cause for concern.

PMS is not directly correlated with any of the classical plant quality indicators, but predawn PMS measurements

Table 7.2.4—*Growth response and cultural implications of inducing moisture stress in conifer seedlings in Northwestern United States nurseries (modified from Landis and others 1989)*

Predawn PMS value (bars)	Moisture stress rating	Seedling response and cultural implications
0 to 5	Slight	Rapid growth
5 to 10	Moderate	Reduced growth Best for hardening
10 to 15	High	Restricted growth Variable hardening may result
15 to 25	Severe	Potential for injury
> 25	Extreme	Injury or mortality

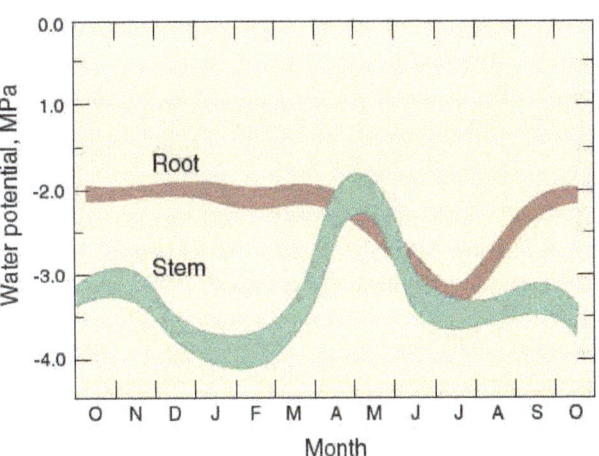

Figure 7.2.12—*The value of water potential at zero turgor varies differently through the year for roots and stems of Douglas-fir seedlings (modified from Ritchie and Shula 1984).*

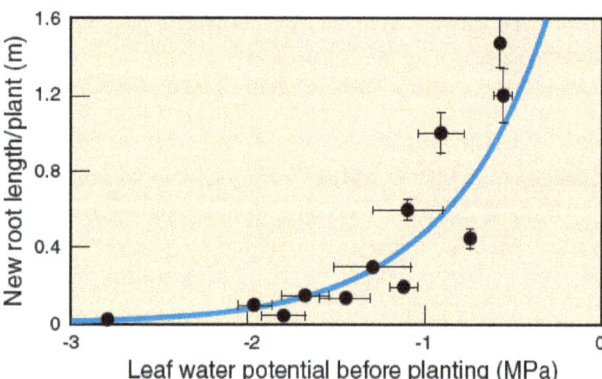

Figure 7.2.13—*In some studies, plant moisture stress was found to be a good predictor of the ability to grow new roots after outplanting (modified from Mena-Petite and others 2001).*

can be used in nurseries to determine irrigation amount and timing, and are the best measurements for monitoring stress levels during hardening. PMS reading during harvesting can alert nursery managers to stressful conditions, and plant users can use PMS to check moisture status of their stock immediately before outplanting.

7.2.4.2 Cold hardiness

Cold hardiness (CH) testing has been used in horticulture since the early 1900s as a method of selecting cold hardy cultivars. Its use as a plant quality test in forest and conservation nurseries has developed over the past 30 or so years, but it stands now as perhaps the second most-often used test of forest planting stock quality.

Concepts behind the test. During the growing season, most temperate zone plants are killed when the air temperature drops below freezing. As winter approaches and growth slows, however, plants respond to the changing photoperiod (lengthening nights) and develop tolerance to cold (Bigras and others 2001; Glerum 1976, 1985; Weiser 1970). In general nursery terminology, this is known as "hardening" and this cold tolerance is indicative of general stress resistance. When winter arrives, plants that would have been killed at slightly below 0 °C (32 °F) during the growing season are able to survive temperatures far below that. As winter draws to a close and the growing season nears, this resistance to low temperatures is rapidly lost and plants resume growth.

What happens when plant tissues freeze? To understand how plants withstand subfreezing temperatures, it is first necessary to understand what happens inside a plant when it freezes. Consider a generalized cross section of plant tissue showing the cellular structure (fig. 7.2.14A). Cells are enclosed by flexible walls made primarily of cellulose, which is stiff and strong. Cells are typically packed tightly together, but occasionally spaces that contain only air and/or water occur between them (intercellular).

Plant tissue is composed of many types of cells that have different functions. Some cells, such as vessels and tracheids, are hollow and transport water from roots to the leaves, or photosynthate back down from leaves. Living cells that function in photosynthesis and other physiological processes are filled with cytoplasm, which is surrounded by a semipermeable membrane composed of a fatty material called lipid in which protein molecules are embedded. This membrane plays a key role in plant cold hardiness; everything within it is referred to as symplast and is living tissue. Everything outside this membrane (cell walls, vessels, intercellular spaces, empty cells, etc.) is referred to as apoplast and is not living (fig. 7.2.14A).

Both the symplast and apoplast normally contain some water. Apoplast water is nearly pure, so its freezing point is close to 0 °C (32 °F). In contrast, the symplast contains dissolved sugars and salts, suspended starch granules, and protein molecules. These solutes act as "antifreeze," depressing the freezing point of the symplast to considerably below 0 °C. So, when cells are exposed to sub-freezing temperatures, the apoplastic water begins to freeze. As it does, small ice crystals form within the cell walls, intercellular spaces, and other voids within the apoplast (fig. 7.2.14B). The symplast water, with its lower freezing point, resists freezing. Therefore, the ice that forms within the plant tissue is contained in the apoplast and does little or no damage.

Ice, however, has a very strong affinity for water—so strong that ice crystals pull water tenaciously across the membrane and out of the symplast. Because the membrane is permeable only to water, the dissolved sugars and other materials remain in the symplast even as water is drawn out. This raises the concentration of the dissolved solutes, further lowering the freezing point of the symplast water. When plant tissues are not cold hardy, or when the temperature falls below the plant tissues' seasonal level of hardiness, the cytoplasm can become severely dehydrated to the point at which: (1) proteins denature; (2) membranes are killed or damaged allowing cell contents to leak into the apoplast; (3) cells plasmolyze; and (4) cytoplasmic cell volume decreases sharply, signaling cell death. It is not clear whether low temperature itself, or desiccation, or both actually incite the damage (Adams and others 1991; Sutinen and others 2001).

Cold injury must be distinguished from winter desiccation that results when cell water is pulled across the cell membrane to feed ice crystals growing outside the cells. This can severely dehydrate cytoplasm and injure membranes causing them to leak cell contents. Even cold hardy plants can be damaged by winter desiccation.

Cold hardiness mechanism. For plants to resist freezing, several changes must occur in the physical and chemical properties of the membranes and the cytoplasm during the hardening process (Öquist and others 2001; Sutinen and others 2001). First, membranes change physically, becoming more permeable to water. This enables water molecules to move out of the cells rapidly, permitting intracellular solute concentrations to increase quickly. In addition, the membranes become physically more rigid. This helps protect them from being pierced by ice crystals that are rapidly growing in the apoplast, while enabling them to resist being torn and pulled away from the cytoplasm and/or cell walls as the cytoplasm dehydrates and shrinks. The cytoplasm itself undergoes profound physical-chemical changes that enable it to survive severe dehydration. These adaptations take place in response to changes in photoperiod and lowering temperatures and are orchestrated by suites of genes that are turned "on" or "off" by these environmental signals.

An important hardiness avoidance mechanism is deep supercooling of water (Burr and others 2001; Quamme 1985). Pure water can cool to nearly –40 °C (–40 °F) without forming ice crystals when no ice nuclei are present, and some plants exploit this property. When supercooled water does freeze, however, it is nearly always lethal. The observation that many plant species do not occur north of the –40 °C midwinter isotherm suggests they avoid cold damage primarily by this mechanism (George and others 1974). This same midwinter isotherm

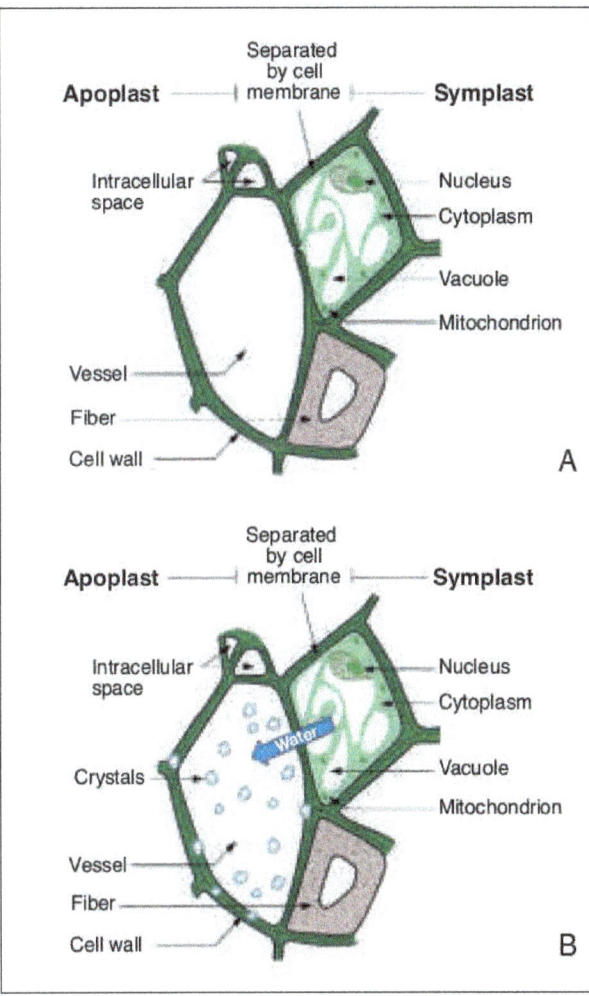

Figure 7.2.14—*Living cell contents (symplast) are separated from nonliving cell contents (apoplast) by the cell membrane (A). When temperatures fall below freezing, ice crystals begin to form in the apoplast. As these crystals grow, they draw water across the cell membrane causing dehydration of the cell contents (B). If the cytoplasm becomes severely dehydrated, the membrane can rupture, and cell contents leak into the apoplast, causing cell injury.*

Table 7.2.5—*Stages of cold hardening and dehardening for coastal Douglas-fir seedlings (compare with figure 7.2.15)*

Hardening stage	Season	Environmental cues	Temperature tolerance as LT_{50}
Hardening begins slowly	Early fall	Shortening photoperiod	–2 to –5 °C (28 to 23 °F)
Hardening increases rapidly	Late fall	Increasing lower temperatures, especially at night	–10 to –20 °C (14 to –4 °F)
Maximum hardiness	Midwinter	Very cold temperatures	–15 to –40 °C (5 to –40 °F)
Dehardening happens quickly	Late winter	Rising temperatures and longer days	Rapidly rising to –2 °C (28 °F)

commonly coincides with timberline, causing Becwar and others (1981) to speculate that supercooling may also limit survival of certain species to below timberline. Many conifers (pines excluded) employ supercooling as a method of avoiding cold damage. However, many tree species can survive temperatures far below –40 °C so they are able to resist cytoplasmic desiccation by other, less well understood, mechanisms.

Stages in cold hardening. Cold hardening (also known as cold acclimation) occurs in a series of stages depending on plant species (Cannell and Sheppard 1982; Timmis 1976; Timmis and Worrall 1975). Table 7.2.5 gives a generalized cold hardening pattern for coastal Douglas-fir shoots and root systems, which is illustrated in figure 7.2.15. The Y-axis represents the LT_{50} value—the temperature that is lethal to 50 percent of a sample population—which is the most common index of cold hardiness.

Further information on the environmental cues that trigger and sustain the various stages of hardening and dehardening are discussed in Greer and others (2001).

Hardiness variation in plant tissues, species, and ecotypes. Different plant tissues harden and deharden at different rates (Bigras and others 2001; Rose and Haase 2002). In particular, the fact that roots do not harden as deeply as shoots (fig. 7.2.15) has very important implications for container growers (Colombo and others 1995). Burr and others (1990) tested cold hardiness of Engelmann spruce seedlings throughout winter and separately examined buds, needles, and lateral cambium (fig. 7.2.16).

Stems and needles hardened more rapidly and achieved greater midwinter hardiness than buds. All three tissues dehardened very rapidly in late winter.

Tree species and ecotypes exhibit a vast range of midwinter hardiness levels depending on the regional climate where they naturally occur (Sakai and Weiser 1973). Boreal conifers, such as black and white spruce, jack pine (*Pinus banksiana*), and others, attain hardiness levels below –80 °C (–112 °F). Many Rocky Mountain conifers, such as lodgepole pine (*Pinus contorta*) and Engelmann spruce, also reach this cold hardiness level. In contrast, Pacific coast conifers, such as Douglas-fir, coast redwood (*Sequoia sempervirens*), and western redcedar (*Thuja plicata*), rarely acclimate to below –20 °C (–22 °F). Note that the cold tolerance of wide-ranging species, such as Douglas-fir, varies by ecotype (–20 °C [–4 °F] for Washington State, but interior sources from the Rocky Mountains can tolerate –20 to –30 °C [–4 to –22 °F]).

Cold hardiness testing methods. Although plants can be tested for cold hardiness by several methods (Burr and others 2001), two tests are widely used: the whole plant freezing test (WPFT) (Tanaka and others 1997) and the freeze-induced electrolyte leakage (FIEL) test (Burr and others 1990; Dexter and others 1932; McKay 1992). Both tests employ two steps (Burr and others 2001; Ritchie 1991). First, plants or plant parts are exposed to a freezing stress and, second, the amount of cold injury is rated. These tests are compared in table 7.2.6.

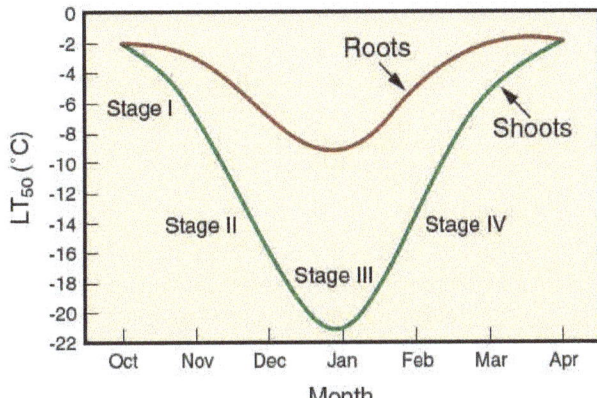

Figure 7.2.15—*These typical cold hardening trends for conifer seedlings show that shoots and roots follow the same general pattern, reaching peak hardiness in January. It is important to note that some species and ecotypes do not reach Stage III Hardening, and roots do not attain the same level of hardiness as shoots.*

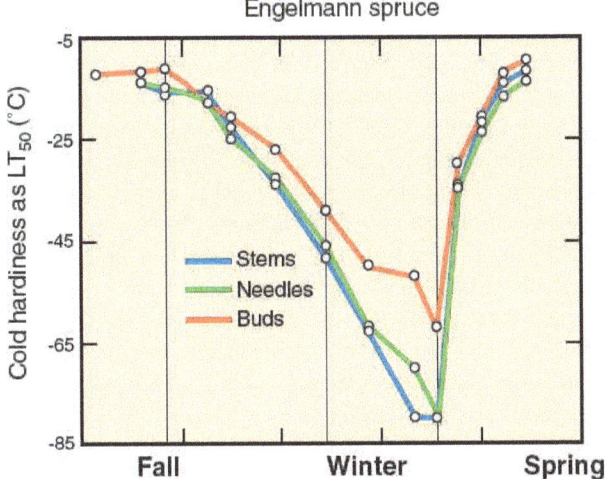

Figure 7.2.16—*Plant tissues harden at different rates in the fall, but all deharden very rapidly in the spring (modified from Burr and others 1990).*

Whole plant freezing test. To begin, a representative sample of plants is subjected to a series of subfreezing temperatures in a programmable chest freezer (fig. 7.2.17A&B) or a Thermotron for a predetermined time period, often a few hours. Next, plants are incubated in a warm environment, such as a greenhouse, for several days to allow symptoms to develop. Finally, the stem, buds, and foliage of test plants are assessed for cold damage by evaluating visible damage or "browning" in the bud, cambial, and foliar tissues (fig. 7.2.17C–E). Mortality is determined based on the severity and position of tissue damage (Tanaka and others 1997).

Freeze-induced electrolyte leakage test. This test is based on the fact that freeze-damaged cell membranes leak electrolytes that can be measured with an electrical conductivity (EC) meter. To begin, sample tissues (foliage, buds, or roots) are cut from the test plants (fig. 7.2.18A), and subjected to freezing temperatures (fig. 7.2.18B). They are then placed into deionized water, which has zero electrical conductivity (fig. 7.2.18C). The electrolytes that leak from damaged cells increase the EC of the water, and this relative increase in EC (described below) is a measure of the amount of cold injury. Although this test can be done on any plant tissue, samples of foliage or roots are most commonly used.

A relative conductivity (RC) index of freeze damage, described by Ritchie (1991) and Burr and others (2001), is determined as follows: (1) place tissue into vials containing deionized water; (2) expose the tissue to sub-freezing temperatures; and (3) incubate the vials until the EC reading stabilizes. This point is known as the initial solution conductivity (EC_1). Finally, the sample is completely killed by heating or freezing and the final conductivity (EC_2) is measured. Relative conductivity is calculated as:

$$RC~(\%) = (EC_1 - B_1) \times 100 / (EC_2 - B_2)$$

where B_1 and B_2 are optional blanks included to account for possible ion leakage from vials.

So, as you can see, the FIEL test provides a quick and easy way to measure cold hardiness of plant tissues.

Differential thermal analysis. Differential thermal analysis (DTA) is based on the theory that when supercooled water

Table 7.2.6—*Comparison of two main cold hardiness tests*

Factor	Whole plant freezing test (WPFT)	Freeze-induced electrolyte leakage (FIEL)
Plant tissue tested	Intact plant (foliage, buds, stem, and roots)	Detailed tissue (foliage, buds, stems, or roots)
Time	Several days to a week	1 to 2 days
Required testing equipment	Programmable freezer and growth chamber or greenhouse	Programmable freezer, electrical conductivity meter, autoclave, oven, or microwave
Evaluation criteria	Degree of tissue damage (browning) or chlorophyll fluorescence (see Section 7.2.4.4)	Numerical reading

freezes it almost always indicates significant tissue injury. Two plant tissue samples (stem or bud) are collected and one is killed by heat or cold and then dried. Two tiny thermocouples, wired in series, are placed in the sample material—one in the dead tissue and one in the living tissue. The samples are placed into a freezing cabinet capable of freezing down to about –40 °C (–40 °F).

As the temperature is slowly lowered, the temperature difference between the samples remains at zero until a freezing event happens. At this point a "spike" is registered. The first spike often occurs when the temperature reaches –5 °C to –10 °C (23° to 14 °F) and represents the freezing of intercellular (apoplastic) water. In tissues that supercool, a second spike will occur at a lower temperature (down to –40 °C [–40 °F]). Evidence suggests that the temperature of this second spike indicates the lethal temperature for that sample (Ritchie 1991).

While this method seems to offer promise for determining hardiness levels of species that supercool, various technical problems have hindered its operational use (Burr and others (2001).

Cold hardiness testing through gene expression. We indicated earlier that changes in environmental signals, specifically photoperiod and temperature, trigger changes in gene expression that ultimately result in cold hardiness development. A novel approach to measuring hardiness, described by Balk and others (2007), involves identifying genes known to be implicated in this process. These genes are responsible for production of enzymes, proteins that trigger all physiological processes in organisms. To create an enzyme, the cell must first transcribe the genetic information stored in the DNA into messenger RNA (mRNA). The strand of mRNA then moves over to a ribosome, a site of protein synthesis, where amino acids are stitched together using the mRNA blueprint. The subsequent chain of amino acids is another enzyme that folds into its characteristic shape, floats free, and begins performing a specific reaction (fig. 7.2.19A). Changes in levels of enzymes triggered by these genes signal acquisition or loss of cold hardiness. An advantage is that these signals can be detected much earlier (indicating that nursery treatments used to trigger cold hardiness development were effective, or that plants are losing cold hardiness in spring) than waiting for measurable changes in cold hardiness values using tests like whole plant freezing and freeze-induced electrolyte leakage.

Research with Scots pine (*Pinus sylvestris*) and Norway spruce (*Picea abies*) identified three indicator genes and their subsequent enzymes that together provide enough information to give an accurate estimate of the cold hardiness stage of nursery plants (Balk and others 2007). Subsequent work with Douglas-fir showed similar results (Balk and others 2008). Chemical assays were developed to detect the enzymes created by the indicator genes, and a private company, N-Sure, now offers this test. A composite sample of bud tissue is collected by the nursery manager, stabilized using chemicals provided in a sampling kit, and

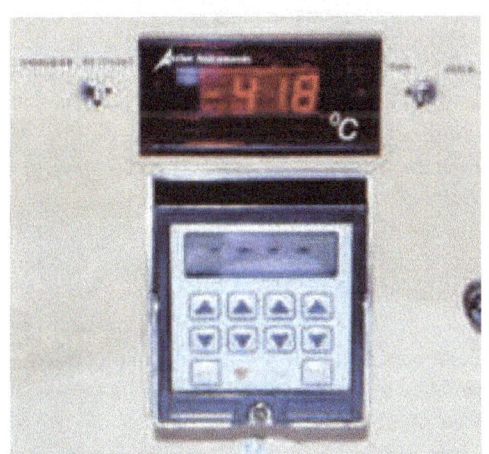

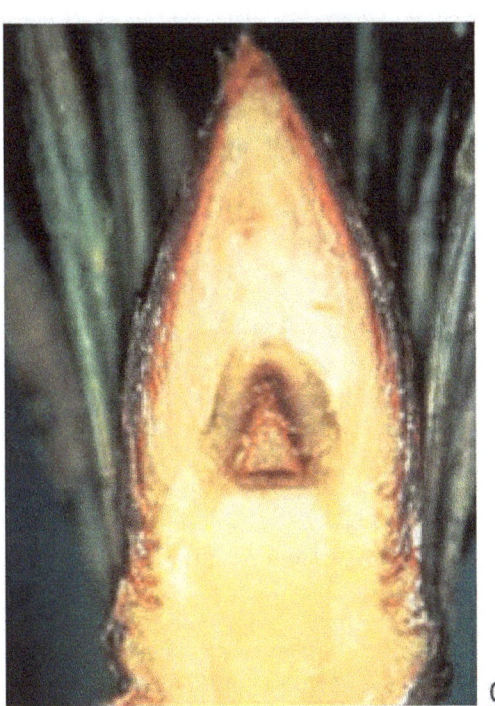

Figure 7.2.17—In the whole plant freezing test, plants are exposed to cold temperatures in a chest freezer (A) with programmable capabilities (B). After a specified exposure period, plant tissues are rated for "browning" of buds (C), foliage (D), and lateral cambium (E).

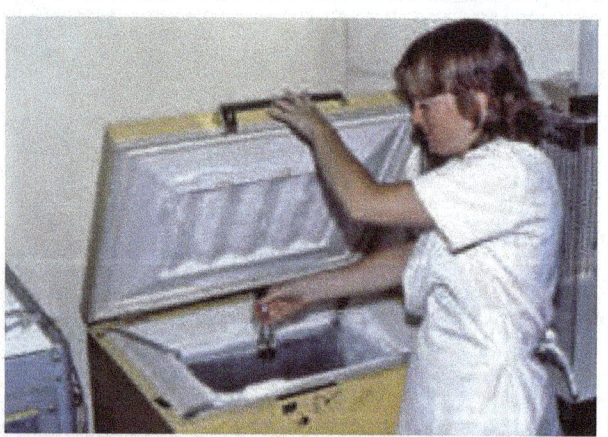

Figure 7.2.18—*In the freeze-induced electrolyte leakage test, plant tissue samples (A) are exposed to freezing temperatures (B) and then immersed in deionized water. The relative increase in electrical conductivity is an indication of cold injury (C) (C, courtesy of Sonia Gellert).*

mailed to the test laboratory (fig. 7.2.19B). Results are available in a few days.

Applications of cold hardiness testing. Container nurseries use CH testing for several purposes.

1. CH tests can be used to track the hardiness of crops as they go through natural hardening in the fall or through cultural hardening procedures, such as blackout. In outdoor compounds, CH tests at regular intervals can be used to determine when frost protection measures are needed (Perry 1998).

2. CH test are commonly used to determine the "lifting window" for container crops. For example, the ability to tolerate -18 °C (0 °F) is being used as an indication of when conifer crops in British Columbia can be lifted for freezer storage (Burdett and Simpson 1984). Different reference temperatures should be developed for other species and ecotypes.

3. CH tests provide a good estimate of overall plant stress resistance (Ritchie 2000), which is a key quality attribute (see Section 7.2.5.2).

Cold hardiness: Summary. Plants that are easily killed by freezing temperatures during the growing season can survive much lower temperatures in winter when they are cold hardy. Cold injury must be distinguished from winter desiccation that results when cell water is pulled across the cell membrane to feed ice crystals growing outside the cells. This can severely dehydrate cytoplasm and injure membranes, causing them to leak cell contents. Even cold hardy plants can be damaged by winter desiccation.

Cold hardening is triggered in late summer by photoperiod and increases during early winter as plants are exposed to increasingly lower temperatures. The level of hardiness can vary greatly among species and ecotypes and is highly influenced by the climate of origin. Peak hardiness occurs in January in temperate zone plants. Following this peak, hardiness can be rapidly lost as plants respond to lengthening photoperiod and warming temperatures.

The most commonly used CH tests are the whole plant freeze test, in which entire plants are exposed to freezing temperatures then evaluated for their response, and the freeze-induced electrolyte leakage test, which is used to test foliar and root samples. Tests based on genetic indicators are now becoming available.

Cold hardiness tests can be used to establish lifting windows, for indicating when frost protection may be needed in the nursery, and as a surrogate for stress resistance testing.

7.2.4.3 Root electrolyte leakage

Roots are among the most fragile parts of plants and, hence, are sensitive to many environmental and operational stresses. This is particularly true of container stock whose root systems are not insulated by surrounding soil. Stresses include high and low temperatures (Lindström and Mattsson 1989; Stattin and others 2000), desiccation (McKay and Milner 2000), rough handling (McKay and White 1997), improper storage (Harper and O'Reilly 2000; McKay 1992; McKay and Mason 1991), and even water logging and disease. It is sometimes possible to detect root damage using the time-honored thumbnail scraping and browning examination, but often the damage is invisible or difficult to quantify. A more rigorous test is called root electrolyte leakage (REL). Because it measures the health and function of root cell membranes, REL can be used as an indication of root injury and, therefore, quality.

REL has been used in Canada (for example, Folk and others 1999) and is currently one of a battery of plant quality tests developed by the Ontario Ministry of Natural Resources (Colombo and others 2001). In the United States, electrolyte leakage has primarily been used to test the cold hardiness of foliage, but application of this technique to roots is uncommon.

The REL method is relatively simple, uses readily available equipment, produces results quickly, and can be useful with deciduous trees, which are leafless in winter (Wilson and Jacobs 2006). Interpretation of REL results, however, can be problematic due to species, seedlot, and seasonal effects.

Theory. REL tests are based on the same principle as the FIEL test described in the previous section. The main difference, however, is that the REL test measures all types of root injury, not just cold damage. The basic idea is that measuring the quantity of ions that leak across damaged

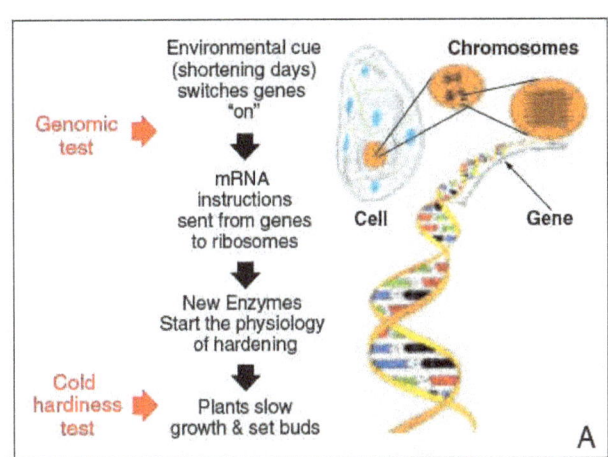

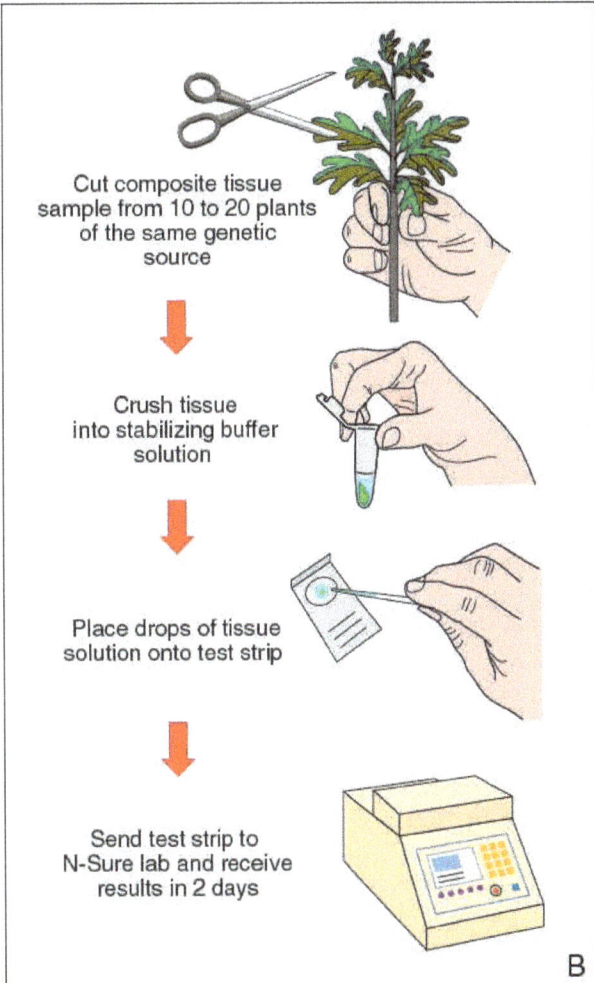

Figure 7.2.19—*Genomic cold hardiness tests allow early detection of the chemical signals that trigger cold hardiness and can serve as an early indicator (A). The N-Sure test provides a quick and accurate way to monitor cold hardiness of nursery stock (B).*

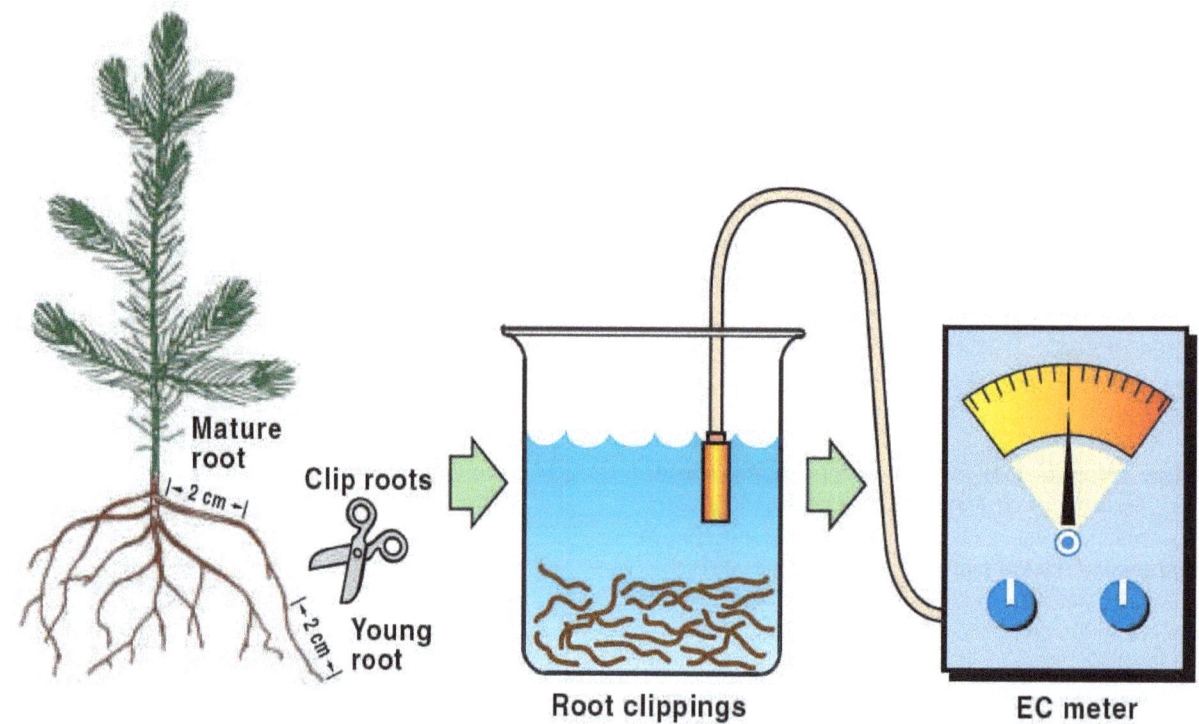

Figure 7.2.20—In the root electrolyte leakage test, measuring the change in electrical conductivity of root tissue gives an indication of the amount of membrane damage. Because this test reflects all types of root injury, it can be used to indicate how well roots will grow out after outplanting.

root membranes provides an estimate of the relative "viability" of the root system (Palta and others 1977). When damaged roots are placed in distilled water, the amount of membrane leakage can be easily and quickly measured with an electrical conductivity (EC) meter.

The biological significance of REL. McKay (1998) offers the following explanation for why the REL test has application as a plant quality test. After outplanting, the main cause of plant mortality is transplant shock induced by water stress. Plants with existing, viable root systems are more efficient in extracting water from soil, and REL measures that root system viability. A low REL reading indicates high root viability, allowing water uptake to mitigate transplant shock.

Measurement procedure. The technique most often used (McKay 1992, 1998) has changed little from the initial protocol described by Wilner (1955, 1960). The steps are as follows (fig. 7.2.20):

1. Roots are first washed in water to remove soil, then in deionized water to remove any surface ions that may be present.

2. A central mass of roots is removed from the plant—with nursery plants this is often a band about 2 cm wide running across the midsection of the root system.

3. Roots with diameter > 2 mm are removed from the sample, leaving only "fine" roots.

4. Fine roots are placed into a vessel containing deionized water.

5. The vessel is then capped, shaken, and left at room temperature for about 24 hours.

6. Conductivity of the solution (C_{live}) is measured with a temperature-compensated EC meter.

7. Root samples are removed and killed by autoclaving for 10 minutes at 100 °C (212 °F) or heating in an oven at 90 °C (194 °F) for 6 hours.

8. Conductivity of the solution surrounding the dead root samples (C_{dead}) is measured.

9. REL is calculated as the ratio of the EC of live roots divided by the EC of dead roots:

$$REL = (C_{live} / C_{dead}) \times 100$$

Applications of REL in nurseries. The REL test is most often used to assess effects of cold damage, poor storage conditions, root exposure causing desiccation, or rough handling of nursery stock. Nearly all the published work has been with bareroot conifer seedlings, primarily Douglas-fir, spruces, pines, and larch. Use of REL to detect freezing damage to roots is applied in one of two contexts: evaluation of cold hardiness test results, and detection of root injury following unseasonably cold weather.

Measuring root cold hardiness. REL cold hardiness testing is the same process as FIEL as explained in Section 7.2.4.2. For example, root samples from bareroot Norway spruce seedlings were exposed to either –5 °C (23 °F) or –10 °C (14 °F) biweekly from September through December in Sweden (Stattin and others 2000). As winter progressed, the difference in REL between cold-treated and nontreated seedlings became smaller, indicating that seedlings were becoming increasingly more cold hardy (fig. 7.2.21).

Detecting cold or heat injury to roots. Because roots of container plants are not protected by the thermal mass of soil, they can be easily injured by extreme temperatures. This is especially true when nursery stock is overwintered outdoors under snow, as is done in eastern Canada and Scandinavia (Lindström and Mattsson 1989). If snow fails to accumulate, or a sudden warm period occurs, container crops are often exposed long enough for their roots to be severely damaged. The REL test is ideally suited for making rapid assessment of potentially damaged nursery stock (for example, Coursolle and others 2000).

Determining lifting windows. REL has been used to indicate when it is safe to harvest bareroot nursery stock (McKay and Mason 1991). For example, Douglas-fir seedlings harvested during midwinter showed much lower REL readings and, therefore, less root injury than stock harvested earlier (fig. 7.2.22).

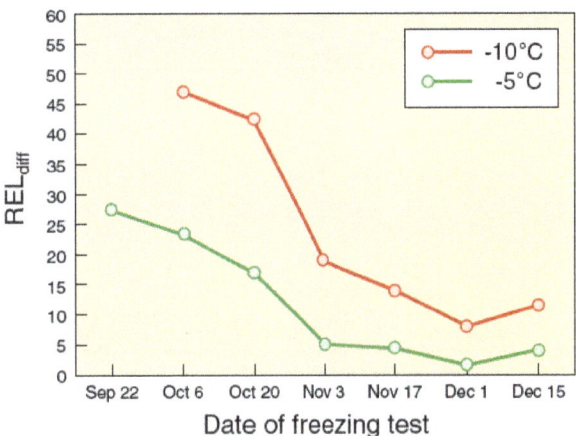

Figure 7.2.21—*Root electrolyte leakage measurements of Norway spruce seedlings show the development of root hardiness during fall. REL_{diff} is the increased electrolyte leakage from roots following exposure to –5 or –10 °C (23 or 14 °F) compared with leakage from nonfrozen seedlings (modified from Stattin and others 2000).*

Monitoring quality of stored seedlings. REL can be used to monitor quality during overwinter storage (McKay 1992, 1998; McKay and Morgan 2001). In one test (McKay 1998), spruce and larch seedlings were harvested throughout winter, beginning October 1, and then placed in storage at 1 °C (33 °F). All seedlings were removed from storage, tested for REL, and then outplanted in April. With both species, REL decreased and survival increased as harvesting was delayed. In another experiment (Harper and O'Reilly 2000), Douglas-fir seedlings were harvested in October, November, December, and January; "warm stored" at 15 °C (59 °F) for 7 and 21 days; and then tested for REL. The REL readings taken at the time of harvest decreased with later harvesting dates, indicating that seedlings were becoming hardier. For each harvest date, however, the readings increased sharply with storage duration suggesting that warm storage contributed to fine root degradation (fig. 7.2.22).

Desiccation and rough-handling effects. Bareroot Sitka spruce and Douglas-fir seedlings were held in controlled environment chambers with their roots exposed to drying conditions for up to 3 hours (McKay and White 1997). The REL readings increased with the intensity of the desiccation treatment, indicating root injury. Injury was confirmed when the desiccation treatments had poor outplanting performance on sites with low spring rainfall in Great Britain.

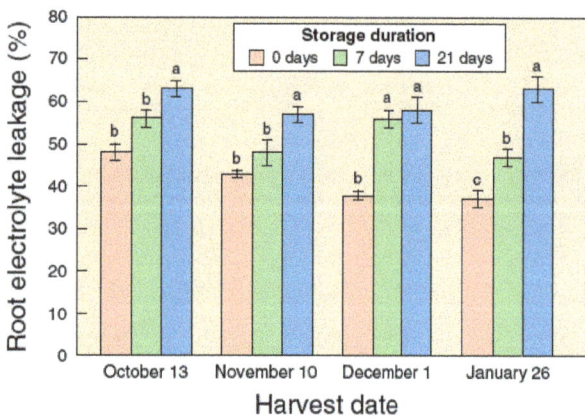

Figure 7.2.22—*REL can be used to determine harvesting (lifting) windows and monitor stock quality during storage. Douglas-fir seedlings harvested during midwinter showed lower REL levels than stock harvested earlier in fall. The same stock was warm-stored after harvesting, and REL measurements at each date showed that less warm storage yielded lower REL levels (modified from Harper and O'Reilly 2000).*

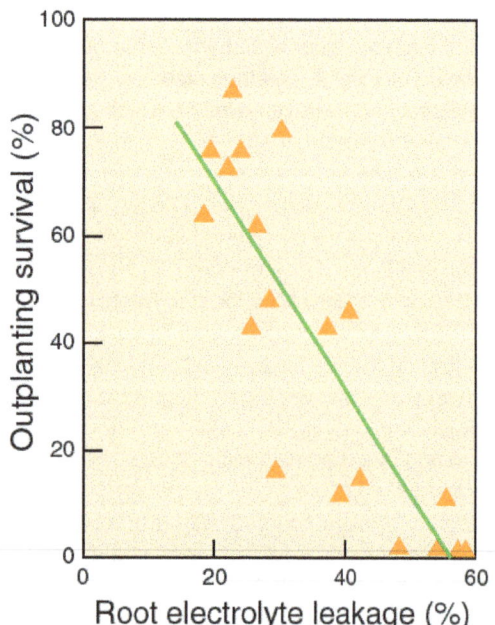

Figure 7.2.23—*Root electrolyte leakage has shown good correlation with outplanting performance in this study with Japanese larch, but not so in many other studies (modified from McKay and Mason 1991).*

Rough handling in combination with root desiccation was assessed in bareroot Douglas-fir, Sitka spruce, Japanese larch (*Larix kaempferi*), and Scots pine using REL (McKay and Milner 2000). Rough-handling treatments were simulated by dropping bags of seedlings from a height of 3 m (9.8 ft). Desiccation was achieved by exposing roots to warm, dry air for 5 hours. Although effects varied with harvesting date and species, REL was significantly higher in stressed seedlings across species and treatments.

REL as a predictor of outplanting performance. The ultimate objective of any plant quality test is to predict how well nursery stock will survive and grow after outplanting, and many studies have used REL for this purpose with mixed results. REL was closely correlated with relative water content of radiata pine (*Pinus radiata*) seedlings 20 days after planting (Mena-Petite and others 2004). With Sitka spruce and Japanese larch seedlings, REL was closely related to both survival and height growth (fig. 7.2.23). In Sitka spruce and Douglas-fir seedlings, REL was correlated with survival on some sites but not others (McKay and White 1997). REL predicted establishment of Japanese larch seedlings to some extent, but root growth potential was a better predictor (McKay and Morgan 2001). Similar results were found with black pine (*Pinus nigra*) (Chiatante and others 2002), while Harper and O'Reilly (2000) reported that REL was a poor predictor of survival potential in warm-stored Douglas-fir seedlings.

Limitations of REL. Why does REL predict survival in some cases but not all? As with many things, "the devil is in the details."

Genetics. REL has been shown to vary with species and even seed sources within species. For example, jack pine and black spruce exposed to a range of damaging root temperatures had REL values in the range of 27 to 31 percent, while white spruce exposed to the same temperatures had REL between 36 and 38 percent (Coursolle and others 2000). Sitka spruce seedlings from Alaska, the Queen Charlotte Islands (QCI), and Oregon provenances were evaluated for their ability to withstand root drying and rough handling (McKay and Milner 2000). Oregon and QCI seedlings exposed to root drying had lower REL values than Alaska seedlings, while Alaska and QCI seedlings, when exposed to rough handling had lower values than Oregon seedlings did. In another study, Douglas-fir had higher REL values than did Sitka spruce, Scots pine, and Japanese larch, regardless of the type of stress encountered (McKay and Milner 2000). Two coastal seedlots of Douglas-fir (British Columbia) gave different relationships between REL and survival (Folk and others 1999).

Dormancy status. McKay and Milner (2000) found that the resistance to stresses mentioned above varied seasonally and was correlated with the intensity of bud dormancy. A similar result was reported by Folk and others (1999) for Douglas-fir seed lots who concluded that REL must first be calibrated to bud dormancy status before it can be effectively used to assess root damage in Douglas-fir.

Seedling age. REL gave good correlations with survival in 2-year-old black pine seedlings, but correlations were weak for 1-year-old seedlings (Chiatante and others 2002). The authors speculate that the efficiency of REL as a quality assessment tool could be closely related to the developmental state of the root system.

Root electrolyte leakage: Summary. Electrolyte leakage from fine roots (REL) is a measure of the ability of membranes within the root system to contain ions. Damaged membranes tend to leak ions so, if ion leakage is quantified, it can provide an indicator of root viability. The REL test is a fast and easy way to evaluate the effects of cold damage, rough handling, desiccation, cold and warm storage, and other stresses on root viability and plant vigor. REL is sometimes closely correlated with plant survival, but in other cases these correlations are weak. This is because factors other than root damage can affect REL, including species, seedlot, plant age, season, and bud dormancy intensity. Fortunately, REL can be calibrated for these effects.

7.2.4.4 Chlorophyll fluorescence

Although technology for measuring chlorophyll fluorescence (CF) has been in place for more than 50 years, it has been applied to tree seedling physiology only since the late 1980s (Mohammed and others 1995). In early trials, forestry researchers considered CF to be an important research tool for potential applications such as assessing effectiveness of irrigation and fertilization, determining harvest windows, and evaluating plant vigor after storage. CF was predicted to be a "simple, rapid, reliable and non-destructive method of evaluating seedling physiological status during the nursery production cycle" (Vidaver and others 1988).

In the intervening years, CF has not lived up to those early expectations. Because CF has such great potential, however, both plant producers and users should have a basic understanding of CF and what it can and cannot do.

What is chlorophyll fluorescence? When solar radiation strikes a leaf, some light energy is reflected, some is transmitted through the leaf tissue, and some is absorbed. Plants absorb much more light energy than is required for photosynthesis. In fact, < 20 percent of the photosynthetically active radiation absorbed by a leaf is actually used in photosynthesis (fig. 7.2.24). Red and blue wavelengths are absorbed by chlorophyll and other pigments, but green wavelengths are reflected, giving living plants their green color. To dissipate all that excess energy that would otherwise be damaging, plants have developed ingenious processes known collectively as "energy quenching." Three types of energy quenching are recognized. Photochemical quenching (qP) is energy used in photosynthesis. Nonphotochemical quenching (qN) is energy dissipated mainly as sensible heat. Fluorescence quenching (qF) is energy emitted as fluorescence and is the basis for the chlorophyll fluorescence test. The largest amount of the absorbed energy is dissipated as sensible heat (qN), while a much smaller amount is given off as fluorescent light (qF) (fig. 7.2.24). These three quenching mechanisms operate simultaneously and in competition with one another.

If these quenching mechanisms are overloaded by high light, the surplus energy drives a biochemical process called the "Moehler reaction." This generates free radicals, mainly oxides and peroxides toxic to the plant. To protect themselves, leaves synthesize scavenging molecules that mop up free radicals and render them harmless. The yellow carotenoid pigments, for example, serve this function. When light intensity is so high, however, as to overwhelm these scavenging systems, then photodamage occurs (Demig-Adams and Adams 1992). This often appears as leaf "scorching" and is common in nursery plants that have been moved too quickly from shade to full sun.

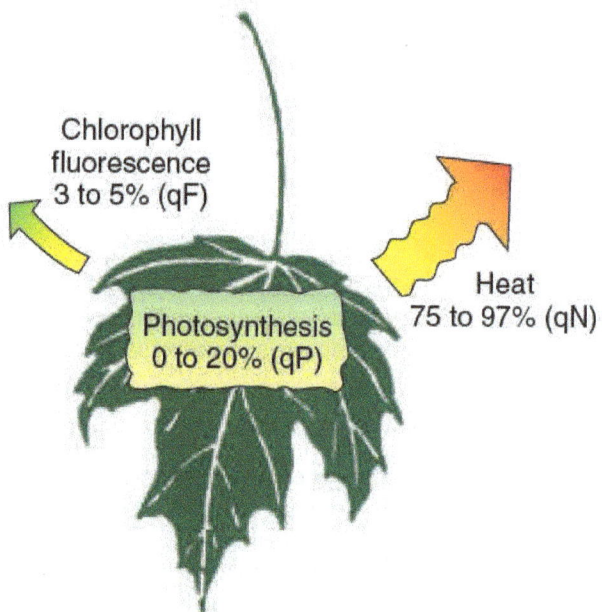

Figure 7.2.24—*Only a small amount of photosynthetically active radiation is absorbed by leaves and actually used (quenched) by photosynthesis. The rest of the surplus energy is quenched as heat loss or as fluorescence.*

The manner in which a plant is able to manage the light energy it absorbs is a sensitive indicator of stress (Krause and Weis 1991). The CF technique, which quantifies energy quenching, is useful for studying plant responses to stress and therefore plant quality.

Photosynthesis and chlorophyll fluorescence. Photosynthesis embodies three sequential processes (Vidaver and others 1991):

1. Light harvesting—light energy is absorbed by light-sensitive pigments (including chlorophyll) in the leaves.

2. Photochemistry—the absorbed light energy is converted into chemical energy.

3. Biochemistry—chemical energy is used to drive Calvin cycle reactions that convert atmospheric carbon into simple sugars.

CF provides a view into the photochemistry process. Because all three processes are intimately interconnected, a perturbation to one part of one process affects the entire set of reactions. These changes in the photosynthetic process are reflected in variations in the amount and rate of CF emissions.

Light energy enters the leaf of a plant and is "captured" by light harvesting pigments (fig. 7.2.25). Depending on the wavelength of the captured light, it enters one of two reaction centers: Photosystem I (PSI) and Photosystem II (PSII), which are located on membranes in the chloroplasts. When a chlorophyll$_a$ (Chl$_a$) molecule in PSII absorbs a photon of energy, one of its electrons is raised to a higher energy state. While in this excited state, it is captured by an electron acceptor pool from which it funnels down through an electron transport chain into PSI, where a similar process occurs (PSI and PSII are named in the order in which they were discovered, not the order of the reaction). This energy transfer leads to the generation of ATP and ultimately the reduction of NADP to NADPH. The energy contained in ATP and the reducing power of NADPH contribute to the fixation of CO_2 molecules and their ultimate conversion to simple sugars in the Calvin Cycle.

"Water splitting" is another key part of the light reaction. In order to replenish the electrons that are lost from Chl$_a$ in PSII, the plant splits water molecules, releasing oxygen atoms into the atmosphere and providing electrons that feed into PSII (fig. 7.2.25).

For any of a number of reasons, many of the excited electrons from Chl$_a$ in PSII are not captured by the acceptor pool and they decay back to their ground state. The energy lost in this decay process is given off as fluorescent light (qF), which emanates entirely from Chl$_a$ in PSII (Krause and Weis 1991) as it decays to its ground state. This is shown in figure 7.2.25 as a wavy line and occurs when the acceptor pool is fully reduced or when the electron transport pathway is backed up. In other words, when more excited electrons are produced than can be processed, they fall back to their ground state, releasing their excitation energy as fluorescence.

This fluorescence emission is too weak to be visible to the naked eye but can easily be detected by an instrument called a chlorophyll fluorometer. The fluorometer measures and quantifies the nature of this fluorescence emission and forms the basis of the CF test.

Measuring chlorophyll fluorescence. The German plant biochemist Hans Kautsky first observed chlorophyll fluorescence in the late 1920s (Govindjee 1995). Kautsky darkened a leaf, then illuminated it with a brief flash of intense

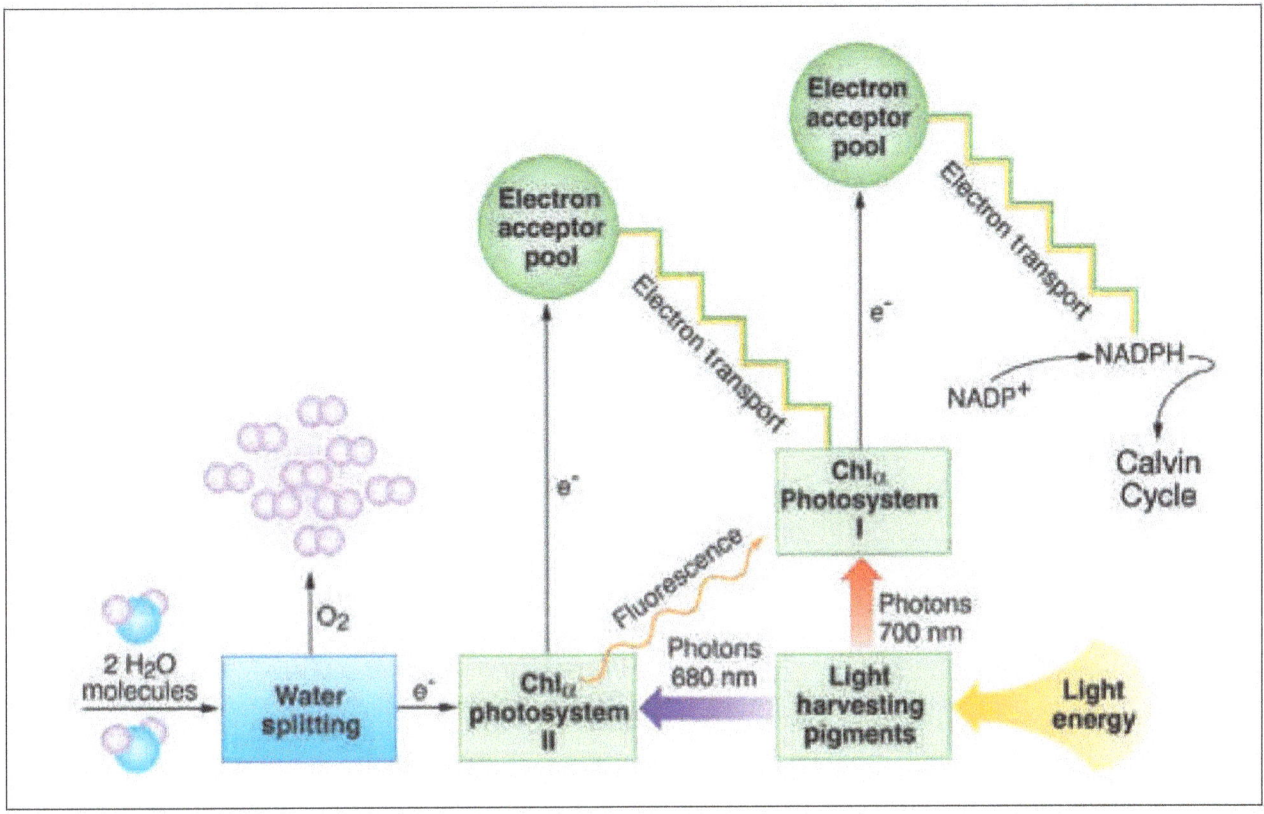

Figure 7.2.25—*Simplified diagram of the light reaction of photosynthesis. Chlorophyll fluorescence emanates from chlorophyll$_\alpha$ in Photosystem II. This fluorescence can be measured with a fluorometer and can be used to diagnose stresses.*

light, and noted an emission of fluorescent light followed the light pulse. Surprisingly, he found that in healthy tissue the emission disappeared within a few minutes, but when the tissue was killed with cyanide or by freezing, the fluorescence emission persisted much longer. It has since been determined that poisoning or freezing leaf tissue disables the electron flow pathway, causing excited electrons to fall back to their ground state and give off measurable fluorescence. In healthy tissue, by contrast, more electrons are quenched in the electron transport pathway, thereby reducing fluorescence emissions.

Kautsky fluorometers. Kautsky's observation led to the development of instruments called "Kautsky" fluorometers. Originally large and cumbersome and the staple of laboratory research on photosynthesis, Kautsky fluorometers have evolved into small, affordable, portable, and user-friendly devices. They contain a light source, two sets of filters, a microprocessor, and a photosensor, and they typically interface with a laptop computer (fig. 7.2.26A). The light source sends a pulse of photosynthetically active light through a fiberoptic cable to the leaf surface where it activates Chl_a in PSII. The Chl_a emission returns back through the cable and passes through a second filter that transmits fluorescent light to the photosensor, which records the emission. The process is controlled by the microprocessor, which is programmed using the laptop computer.

The CF measurement process begins with "dark adapting" the leaf for about 20 minutes. This ensures that: (1) all chlorophyll is in an unexcited, or ground, state; (2) the acceptor pools are empty; and (3) the electron transport pathway is clear before the light pulse is received. Following the light pulse, the fluorometer generates a curve in which the intensity of the resulting fluorescence emission is plotted over time (fig. 7.2.26B). In the Kautsky curve, Fo is fluorescence emanating from the light harvesting pigments in the leaf, not from PSII. Fm is the maximum fluorescence, and Fv is the variable fluorescence coming from PSII.

This curve has many diagnostic features, but the most useful is the ratio of variable fluorescence to maximum fluorescence, or Fv/Fm. This ratio provides a direct estimate of the efficiency of the light reaction (Genty and others 1989) and is the most often used CF output.

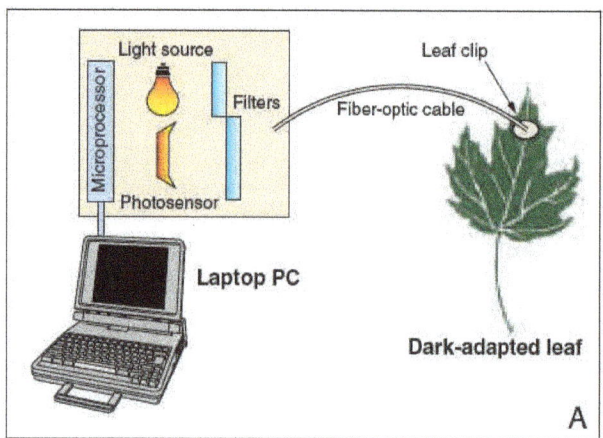

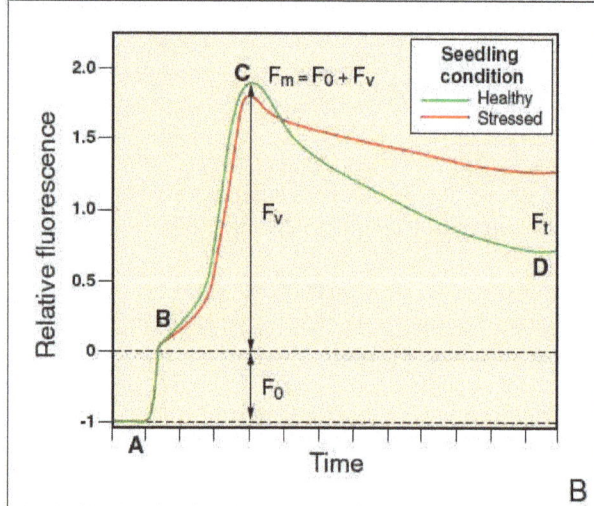

Figure 7.2.26—*A Kautsky fluorometer consists of a light source, two filters, a photosensor, microprocessor, and a fiberoptic cable that attaches to a leaf. Instructions are sent to the fluorometer from a laptop computer (A). A quenching curve is generated after a light pulse is delivered to a dark-adapted leaf. These curves are diagnostic because healthy and stressed plants differ in the amount and duration of their fluorescence emission (B). For example, the ratio of variable to maximum chlorophyll fluorescence (Fv/Fm) is a good indicator of photosynthetic efficiency. See table 7.2.7 for explanation of symbols (B, modified from Rose and Haase 2002).*

Pulse amplitude modulated fluorometers. A more recent development in fluorometry is an instrument called the pulse amplitude modulated (PAM) fluorometer (Schreiber and others 1995). After delivering an initial excitation light pulse, the PAM generates a rapid stream of high-intensity, saturating light pulses that overwhelm acceptor pools, thus canceling out photochemical quenching. The fluorescence emission differentiates between these peaks and the fluorescence decay curve is, therefore, nonphotochemical quenching.

This powerful procedure enables simultaneous measurement of the three energy-quenching components, along with determination of overall process efficiency at several levels. One of these instruments, the PAM-2000, is manufactured by Heinz Walz in Germany (http://www.walz.com). PAMs have become an essential tool for seedling physiology research. A PAM-2000 run produces estimates of quantum yield (Fv/Fm), effective quantum yield (Y), photochemical quenching (qP), nonphotochemical quenching (qN), electron transport rate (ETR), and many other variables.

Normal values of CF parameters in plants. The biochemistry of photosynthesis is essentially uniform across all species of C_3 plants. Therefore, CF parameters in "normal" healthy plants would not be expected to vary across a broad range of species. Discussions with other scientists, as well as perusal of the CF literature, led to the development of table 7.2.7. This gives what are often considered to be "normal" values for the CF parameters and can be used as a guide to interpreting literature values.

Use of CF in plant-quality assessment. At the present time, CF is primarily a research tool but is beginning to be used operationally in some nurseries.

Dormancy. Although attempts to use CF as an indicator of plant phenological condition or dormancy status have been done, we are not yet convinced that these studies are verifiable or repeatable.

Cold hardiness. Currently, the most common use of CF is in detecting and assessing cold injury (Binder and others 1997). For example, when 17 species of *Abies* were tested for cold hardiness, the damage to buds, foliage, and lateral cambium were all well correlated with CF ratings (Jones and Cregg 2006). When compared with other cold hardiness tests, CF was shown to be a quick, nondestructive indication of cold injury of the foliage and stems of Scots pine container stock (Peguero-Pina and others 2008). Rather than rating cold injury with visual, electrolytic, or other methods (see Section 7.2.4.2), the CF approach uses the response of the photosynthetic process as an index of damage. "Normal" plants will typically have Fv/Fm values from 0.700 to 0.830, or slightly lower in winter. When this value falls to < 0.600 following freezing, it indicates significant damage to the photosynthetic process (table 7.2.7).

Outplanting performance. Some studies have attempted to correlate CF variables with outplanting performance. For example, measures of effective quantum yield predicted variations in survival and plant health of stored and non-stored Douglas-fir seedlings in an Irish nursery (Perks and others 2001).

Storage effects. Short-term (2-week) cooler storage of radiata pine seedlings caused depressions of Fv/Fm, Fv/Fo, and other CF parameters as leaf water potential, stomatal conductance, and net photosynthesis also dropped (Mena-Petite and others 2003). These reflected storage-related damage to photosynthetic apparatus and portended reduced post-planting performance. CF is being used as a plant-quality test after storage in some Ontario nurseries (Colombo 2009).

Drought stress. Long-term drought affects photosynthesis directly by depressing leaf water potential, which closes stomata. Recent evidence suggests that prolonged drought also disrupts photosynthesis at the photochemical level. When white spruce seedlings were exposed to 21 successive days without water in a controlled environment chamber (Bigras 2005), Fo and qN were unaffected, but Fm, Fv, Fv/Fm, and qP were depressed when water potential fell below −1.0 MPa (10 bars PMS). Fv/Fm measured in dormant Norway spruce seedlings was unaffected by 4 weeks of post-planting drought in the field, but the same drought exposure depressed Fv/Fm from 0.83 to about 0.28 in seedlings lacking bud dormancy (Helenius and others 2005).

Chlorophyll fluorescence: Summary. Plants have evolved intricate mechanisms for dissipating, or quenching, the light energy they absorb. Some of this energy is used in photosynthesis (photochemical quenching, qP), while the remainder is dissipated by nonphotochemical (qN) or fluorescence (qF) quenching.

Stress caused by high and low temperature, disease, drought, inadequate nutrition, and so on impairs a plant's ability to manage energy quenching. Thus, by measuring and interpreting the three components of quenching with CF, it is possible to detect damage resulting from subtle, transient stress as well as long-term, severe stress. Three important CF parameters that are often reported in the nursery literature are qP, qN, and Fv/Fm.

Damaged or stressed plants have the ability to recover quickly, so it is important to measure CF parameters over a course of several days following stressful events before conclusions about plant damage can be reached. If Fv/Fm remains low and qN high for several days, this indicates that significant damage to the photosynthetic system has probably occurred. Still, much more research is needed before CF will be an operational quality test.

7.2.4.5 Mineral nutrient content

Intuitively, the amount of mineral nutrients that is stored in a plant should be related to its quality. Mineral nutrients such as nitrogen and phosphorus supply the building materials for new growth, and newly outplanted seedlings must rely on a supply of stored nutrients until they are established in the field. Because they reflect actual mineral nutrient uptake, plant tissue tests are the best way to monitor a fertilization program. Analytical laboratories are able to accurately and precisely measure the levels of all 13 mineral nutrients in a small sample of plant tissue, and nursery managers can obtain results in as little as a week. By also measuring tissue biomass, nutrient content can be calculated from the laboratory results for nutrient concentration. That data can then be examined using vector diagrams for relative differences among fertilizer regimes for nutrient dilution, toxicity, sufficiency, or deficiency (Haase and Rose 1995). Although tentative guidelines for analyzing mineral nutrient levels exist, they are for general classes such as "conifer seedlings" (table 7.2.8) and are of limited usefulness for precision monitoring of fertilizer programs. Most published test results are for commercial tree species and almost nothing is known about other native plant species (Landis and others 2005).

Another problem is that correlation between foliar nutrient levels and outplanting survival is not good. One problem is that a plant could be severely stressed or even dead and

Table 7.2.7—*Normal ranges of chlorophyll fluorescense emissions parameters in C4 plants (extracted from the literature)*

CF parameter	Definition	Description	Normal range	Stress range
Fo	Ground state fluorescence	Fluorescence which emanates from the light-harvesting pigments of the leaf; generally considered a "background level" fluorescence that is zeroed out when measuring PSII chlorophyll fluorescence.	0.2 to 0.4	> 0.7
Fs	Steady-state fluorescence	Fluorescence level (sometimes referred to as Ft)		Low Ft indicates stress
Fv	Variable fluorescence	Height of the fluorescence peak above Fo following exposure to the actinic light pulse (Fv = Fm – Fo)		
Fm	Maximal fluorescence	Fv + Fo	1.2 to 1.5	
Fv/Fm	Maximum quantum yield	An estimate of the ratio of moles of carbon fixed per mole of light energy absorbed (Genty and others 1989); theoretical maximum value for C_3 photosynthesis is approximately 0.830.	0.70 to 0.83	< 0.60
Y	Effective quantum yield	(Fm-Fs)/Fs	0.40 to 0.60	0.10 to 0.20
qN	Nonphotochemical quenching	Dissipation of absorbed light energy by means other than photosynthesis (mainly as sensible heat)	0.4 to 0.6	Prolonged values > 0.6
qP	Photochemical quenching	Use of absorbed light energy via photosynthesis	0.7 to 0.8	Prolonged values < 0.6
ETR (in full sun)	Electron transport rate	Speed at which electrons are transported through the photosystem	< 300 µmol electrons $m^{-2}s^{-1}$	

still contain ideal mineral nutrient levels. Even though mineral nutrient levels are not a guarantee of vitality, foliar nitrogen levels appear to be a good predictor of growth after outplanting (Landis 1985). For example, van den Driessche (1984) found a strong correlation between foliar nitrogen and the shoot growth of Sitka spruce seedlings when measured 3 years after outplanting (fig. 7.2.27A). This makes sense only because, after a plant is established, it needs good reserves of nitrogen to repair any injuries and build new cells. Some nurseries have established foliar nitrogen targets at the time of harvest as one indication of plant quality; for instance, provincial nurseries in Quebec specify a minimum foliar nitrogen level for their nursery stock depending on container size (Government of

Quebec 2007). Therefore, the best recommendation is for nurseries to develop their own foliar nutrient standards for the plant species that they grow.

The latest research into the relationship between seedling nutrient levels and outplanting performance involves a concept called "nutrient loading" with nitrogen. The idea is that "supercharging" a seedling with nitrogen will help it survive and grow better on the outplanting site where mineral nutrients are usually limiting. Nutrient loading involves fertilizing seedlings during the hardening phase until their nitrogen content is in the luxury consumption area of the growth curves (fig. 7.2.27B). This process has been successful with black spruce (*Picea mariana*) on sites with heavy plant competition as chronicled by Timmer and his associates (for example, Timmer 1997). The concept of nutrient loading with nitrogen is certainly attractive and it is hoped that this technique will be tested with more species on a wide variety of outplanting sites (Landis and others 2005). Possible problems with increased animal predation and lower frost hardiness also need to be investigated.

7.2.4.6 Carbohydrate reserves

It seems logical that the amount of food stored as carbohydrates in nursery stock should be a good indication of plant quality. After outplanting, nursery plants must rely on this stored "food" to fuel new growth until the plants can start photosynthesizing. Marshall (1983) gives an excellent review of carbohydrates in plants and presents a good comparison of how stored carbohydrates would be used in two different seedlings. Seedling 1 contains adequate levels when harvested, but carbohydrates are gradually consumed during storage; after outplanting, even more are used until the plant becomes established and generates new carbohydrates through photosynthesis (fig. 7.2.28A). Plants that suffered stress or injury would use even more carbohydrates to repair tissues and fuel metabolic recovery. In fact, carbohydrate reserves were found to influence the growth of nursery stock for up to 2 years after outplanting (Ronco 1973).

Unfortunately, research trials have not shown carbohydrate reserves to be a good predictor of plant quality and little has been done with container nursery stock. For example, the carbohydrate reserves of bareroot Scots pine seedlings were evaluated as an indicator of stock quality, and the results followed the general trend in figure 7.2.28A. When reserves dropped below 2 percent total glucose during storage, significant mortality occurred (fig. 7.2.28B). The author concluded that difficulties in measuring carbohydrate concentrations and the dynamics of carbohydrate metabolism make tests of carbohydrate reserves impractical for operational use as a plant-quality index (Puttonen 1986).

Measuring performance attributes can be thought of as a "bioassay" that integrates the functioning of all plant systems into one performance variable. Although they are often robust indicators of plant performance potential, performance attributes do not identify what, specifically, is wrong when performance potential is low. In addition, they also suffer from being very time consuming to measure directly, which can limit their usefulness to plant producers and users.

Table 7.2.8—*Target concentrations for the essential mineral nutrients in the foliage of conifer nursery stock (modified from Landis 1985)*

Nutrient	Symbol	Acceptable range
Macronutrients (%)		
Nitrogen	N	1.30 to 3.50
Phosphorus	P	0.20 to 0.60
Potassium	K	0.70 to 2.50
Calcium	Ca	0.30 to 1.00
Magnesium	Mg	0.10 to 0.30
Sulfur	S	0.10 to 0.20
Micronutrients (ppm)		
Iron	Fe	40 to 200
Manganese	Mn	100 to 250
Zinc	Zn	30 to 150
Copper	Cu	4 to 20
Boron	B	20 to 100
Molybdenum	Mo	0.25 to 5.00
Chloride	Cl	10 to 3,000

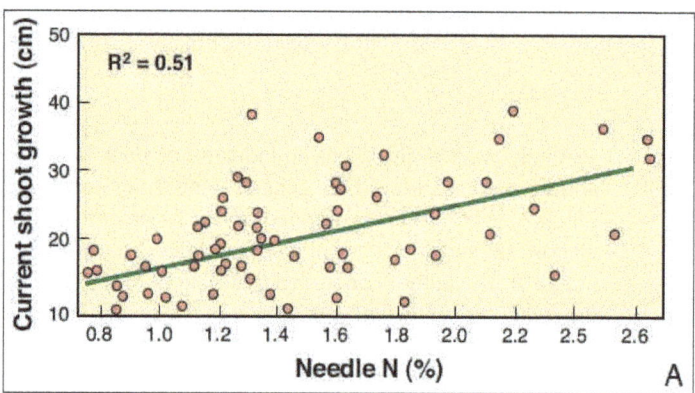
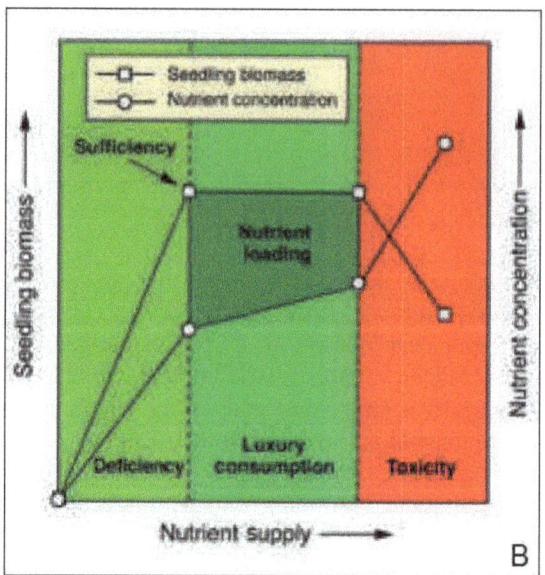

Figure 7.2.27—*Foliar nitrogen (N) concentration was shown to be a good predictor of the shoot growth of Sitka spruce seedlings when measure 3 years after outplanting (A). "Nutrient loading" conifer seedlings with high levels of nitrogen (B) has been shown to be beneficial on wet outplanting sites with heavy plant competition (A, modified from van den Driessche 1984; B, modified from Timmer 1997).*

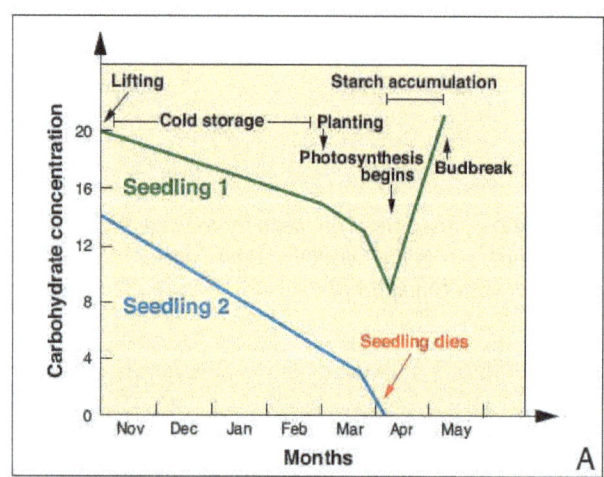
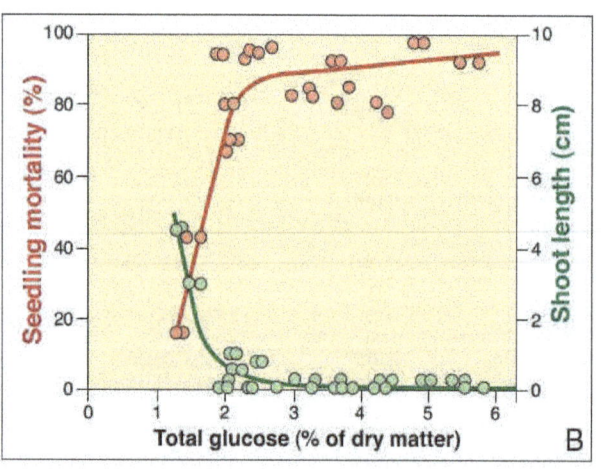

Figure 7.2.28—*Nursery plants consume significant amounts of stored carbohydrates from harvest through storage and outplanting. Seedling 1 contained adequate reserves and survived until it became established on the outplanting site and replenished carbohydrates through photosynthesis. Seedling 2 started out with inadequate carbohydrate storage and died soon after outplanting (A). With Pinus sylvestris seedlings, mortality increased and shoot growth decreased after outplanting when total glucose levels dropped below 2 percent (B) (A, modified from Marshall 1983; B, modified from Puttonen 1986).*

51

7.2.5 Performance Attributes

7.2.5.1 Bud dormancy

The notion that nursery stock quality is related to its dormancy status is strongly ingrained in the minds of plant producers and users, especially foresters. When pressed to explain this relationship and why it is important, however, few are able to articulate a clear view of what dormancy is, how it works, or how it affects quality. So, our intent is to discuss this important concept with the caveat that dormancy intensity can vary between species and ecotypes. In particular, plants from higher latitudes and elevations will show stronger dormancy than those from lower latitudes and elevations.

The concept of dormancy. Dormancy is one of the oldest concepts in plant science. Nursery workers learned by trial and error that plants could be transplanted and outplanted most successfully when they were not actively growing. In the temperate zone, this occurs in winter, so nurseries have traditionally harvested stock then. The concept of the "lifting window" was developed by harvesting and outplanting seedlings from late fall through early spring and measuring survival and growth (Jenkinson and others 1993). These trials supported the traditional practice of harvesting during midwinter, and people interpreted these results to mean that plants were most "dormant" during this period. As we will show, however, this concept of a midwinter dormancy peak is not correct.

Defining dormancy. Dormancy can be broadly defined as a state of minimal metabolic activity, or any time that a plant tissue is predisposed to grow, but does not (Lavender 1984). In other words, dormancy is that condition in which plant growth—cell division and enlargement—is not occurring. In horticulture, dormancy can refer either to seed dormancy or plant dormancy. In the published literature, plant dormancy has been studied much less than seed dormancy but plant dormancy is what we are concerned with here.

Two kinds of plant dormancy are recognized:

> External dormancy, also known as "quiescence," occurs when environmental conditions (for example, severe water stress) will not support growth (Lavender 1984). Plants exhibiting imposed dormancy will resume growth when these unfavorable conditions improve (when it rains).

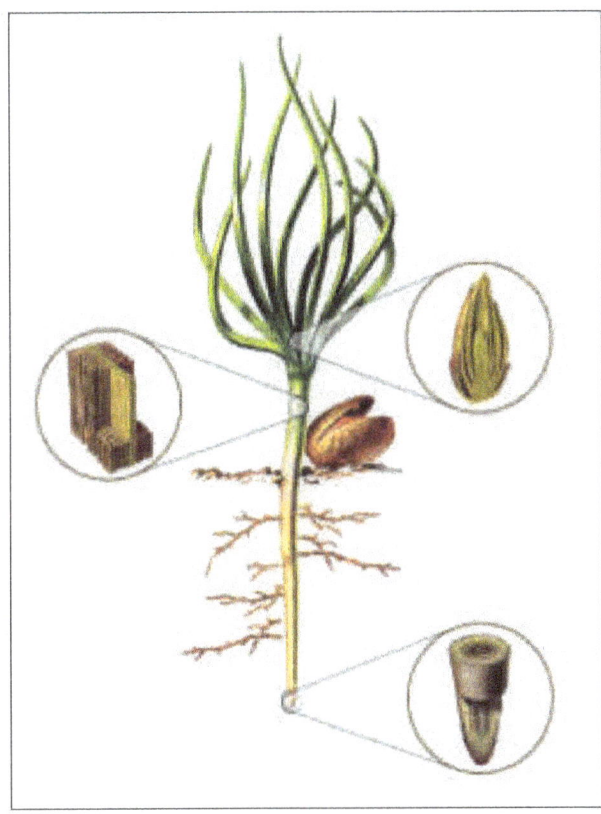

Figure 7.2.29—*Dormancy refers to the activity of the meristematic tissues: buds, lateral meristems in the stem, and root tips. In the normal context of plant quality, bud dormancy is the primary concern.*

> Internal dormancy, or "deep dormancy," is a condition in which plants will not resume growth until they have experienced a long period of exposure to low temperatures (Perry 1971). This condition is also called "winter rest." In this chapter, we are concerned with deep dormancy and how this physiological condition affects nursery culture and outplanting success.

Dormancy refers to tissues, not entire plants. In everyday nursery jargon we talk about plants, or even entire crops, being dormant. While this is common terminology, it is important to understand that plant dormancy refers to a specific meristematic tissue, usually buds (fig. 7.2.29). In the same plant, buds may be dormant while the lateral meristem may not. Root meristems never truly go dormant and will grow anytime that environmental conditions, especially temperature, are favorable. Because we are

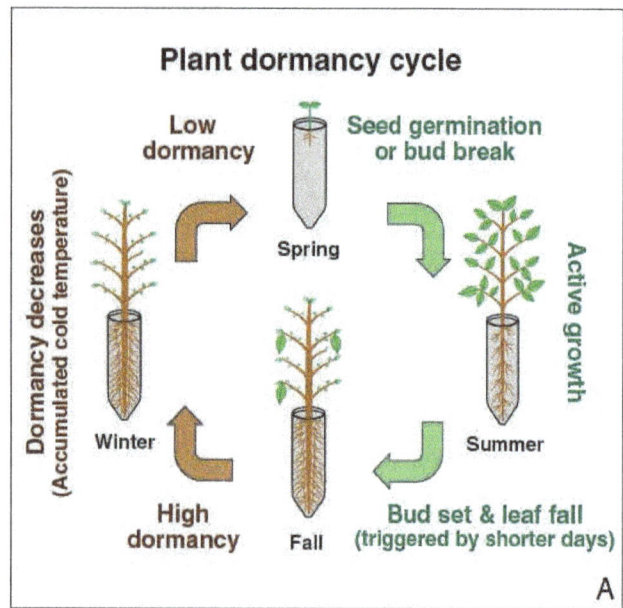

Figure 7.2.30—*The buds of perennial plants in the temperate zone, including forest and conservation nursery stock, undergo a seasonal cycle of shoot growth and dormancy. Note that peak dormancy occurs in late fall instead of midwinter, as is often believed, and that dormancy is released by cumulative exposure to cold ("chilling requirement"). Some dormant plants exhibit morphological changes: firm "winter buds" and bluish needles due to waxy deposits (B), and purplish foliage in others species. Due to extreme variation among individuals (C), these color changes cannot be used to predict dormancy.*

concerned with quality testing, we will be discussing bud dormancy, which is most clearly observed in the behavior of terminal buds.

The dormancy cycle. Perennial plants that grow in temperate regions exhibit a pronounced seasonal "cycle of dormancy" (fig. 7.2.30A). In spring, as day length and temperature increase, plant buds begin to exhibit dimensional increases reflecting both cell division and expansion—in other words, they begin to grow. Shoot growth persists through spring and into summer. In summer, as day length (photoperiod) begins to shorten, the increasing length of the dark period is perceived by the phyto-chrome system in leaves as a signal to begin preparing for winter. At this point shoot growth slows and winter bud development proceeds (Burr 1990). By early fall, some plants form a dormant bud and exhibit other morphological changes, such as leaf color change and abscission in hardwood stock (fig. 7.2.30A), increased needle waxes on conifer needles (fig. 7.2.30B), and purplish needle color in other plants. These visual changes should not be considered proof of dormancy, however, as considerable variation occurs among individuals in the same seedlot (fig. 7.2.30C). In a study with Scots pine seedlings, no predictive relationship could be developed between purplish foliage and cold hardiness test results (Toivonen and others 1991).

The chilling requirement. In late summer, plant buds enter the condition of imposed dormancy. As summer surrenders to autumn, imposed dormancy gradually gives way to deep dormancy and buds reach maximum dormancy in late fall (fig. 7.2.30A). As we just mentioned, dormancy is then released by exposure of the plants to an extended period of low temperatures; this is known as a "chilling requirement" and is sensed by the buds. This evolutionary adaptation ensures that plants will not resume shoot growth (break bud) during a midwinter warm spell only to be killed by a return of cold weather. Once this chilling requirement is satisfied, warm spring temperatures and, to a lesser extent, lengthening photoperiod, will trigger and sustain a resumption of shoot growth (Campbell 1978). Although temperatures in the range of about 3 to 5 °C (37 to 41 °F) are most efficient at releasing bud dormancy (Anderson and Seeley 1993), temperatures above and below this range also are effective to a lesser degree (fig. 7.2.31).

Orchardists and other horticulturalists have developed elaborate models to predict the date of flower bud opening in cold-sensitive crops such as peaches (see, for example, Richardson and others 1974). These models take into account the efficiency of chilling and the fact that warm interruptions during late fall can negate some chilling that has occurred up to that time. In forest and conservation nurseries, however, a simpler process for calculating chill sums or chilling hours is often used. The details are given in the following section.

Measuring dormancy. Because of the tremendous importance of measuring dormancy to nursery management, many attempts have been made to develop a simple way to measure it. As we will now discuss, this objective has been elusive.

Dormancy meters. In the 1970s, researchers observed that changes in electrical resistance of plant tissue provided a useful way to determine whether tissues were injured or dead. Building on these observations, they constructed a "dormancy meter" (fig. 7.2.32) with the objectives of measuring dormancy in fall and telling nursery managers when it was safe to harvest their stock. Unfortunately, subsequent tests showed that these meters were unreliable (Timmis and others 1981). The idea of a simple "black box" quality test is still attractive, but it is doubtful that any equipment or technique will be able to instantaneously measure bud dormancy.

Chilling sums. This is the easiest and most practical method for estimating the intensity of bud dormancy and is based on the chilling requirement just discussed. Chilling sums have immediate application, because they can be used to establish harvesting windows or monitor bud dormancy as it weakens during winter. The concept is logical enough: the cumulative exposure of plants to cold temperatures controls the release of dormancy. So, by measuring the duration of this exposure, it is possible to estimate the intensity of dormancy indirectly.

In actual practice, chilling hours, or degree-hardening-days (DHD), have been used. The process involves meas-

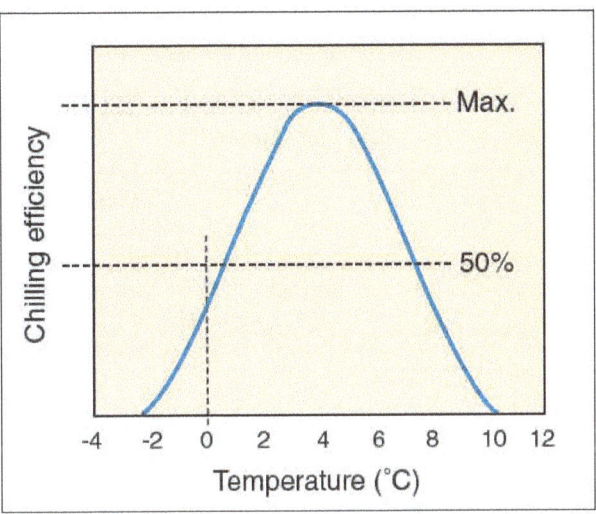

Figure 7.2.31—*Chilling temperatures and their efficiency at breaking bud dormancy (modified from Anderson and Seeley 1993). Note that temperatures in the range of refrigerated storage (−1 to +1 °C [30 to 33 °F]) release dormancy very slowly.*

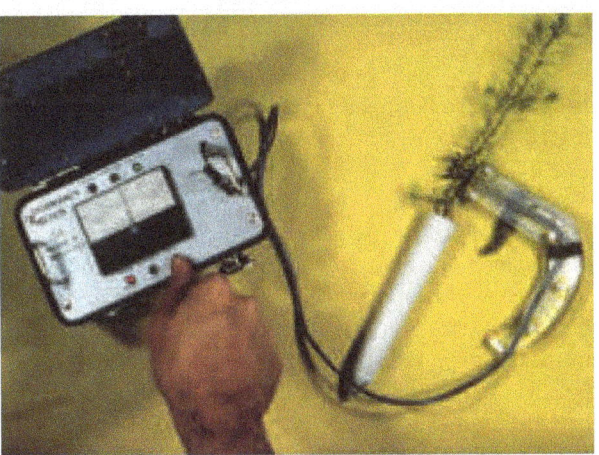

Figure 7.2.32—*The "dormancy meter" was an attempt to find a simple and easy way to measure dormancy and to determine when plants were ready for harvest. Operational testing showed that such devices were unreliable.*

Table 7.2.9—*An example of how to calculate chilling sums using degree days, calculated from an average of daily maximum and minimum temperatures and a 40 °F (4.5 °C) base temperature*

Day	Base temperature (°F)	Daily temperatures (°F)			Degree days	Chilling sum
		Maximum	Minimum	Average		
One	40	40	20	30	10	10
Two	40	45	35	40	0	10
Three	40	50	40	45	0	10
Four	40	40	30	35	5	15

uring the temperature each day and calculating the amount of time below a specific reference temperature. A method sometimes used in forest and conservation nurseries is to simply count the number of hours during which the air temperature is at or below a threshold value, such as 5 °C (41 °F) (Ritchie and others 1985). Reference temperatures will vary with nursery location and species; for example, 8 °C (46 °F) has been used for southern pines (Grossnicle 2008). One shortcut method is to record the daily maximum and minimum temperatures, average them, and subtract this average from the base temperature. Note that, when calculating chilling sums, only negative values are recorded (table 7.2.9).

Bud break test. The more dormant a plant is, the more slowly the terminal buds will resume growth (break) under ideal growing conditions. This phenomenon forms the basis of the only direct way of measuring dormancy intensity—the bud break test. With access to a greenhouse or other growth-promoting structure that can maintain ideal growing conditions through the winter, the intensity of dormancy in nursery stock can be measured by observing days to bud break (DBB) in this "forcing" environment.

The procedure is relatively simple. Grow plants to shippable size and, in the late summer, harden them to the fully dormant condition by exposing them to ambient conditions. By early fall, plants typically have formed a dormant bud and exhibit the other morphological changes, such as leaf color change and abscission in hardwood stock (fig. 7.2.30A) and increased needle waxes on conifer needles (fig. 7.2.30B). Place a temperature recording device at plant height and check temperatures at least weekly to compute chilling sums (table 7.2.9).

Set the environmental controls in the testing greenhouse to maintain spring forcing conditions with warm days, cool nights, and long photoperiods created with photoperiod lights. Then, beginning around Halloween, harvest a sample of plants, pot and label them, and bring them into the forcing greenhouse. Keep the sample plants watered and count the number of days required for the terminal buds to resume growth—this is DBB. Repeat this process at every major holiday: Thanksgiving (late November), Christmas (late December), New Year's Day (early January), Valentine's Day (mid-February), and St. Patrick's Day (mid-March). Starting at the first sample date in September, keep track of the sum of chill hours, all hours when the temperature was, say, 5 °C (41 °F) or lower throughout this test period.

When finished, plot the DBB values over the chilling sums. The number of days required for the terminal buds to break is a direct measure of dormancy intensity. (Note: the Halloween plants may never break bud.) It is likely results will be similar to those shown in figure 7.2.33, which came from coastal Douglas-fir in western Washington and Oregon (Ritchie 1984a) and are in agreement with the general curve proposed by Lavender (1984). As the chilling sum accumulates during winter, the DBB will shorten dramatically. Similar experiments with many tree species, including several hardwoods (birch, dogwood, hawthorn, and oak) have yielded similar results

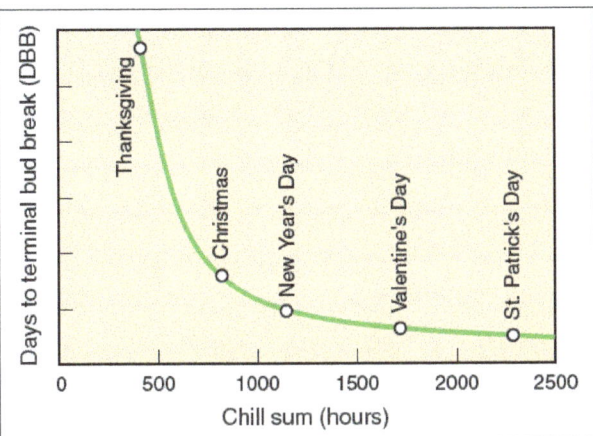

Figure 7.2.33—*The only reliable test for bud dormancy intensity is a bud break test that can be performed by harvesting plants at regular intervals during late fall and winter and bringing them into a greenhouse. As they break bud, the number of days to bud break (DBB) is plotted against the chilling sum for each lift date. The data shown are typical of Douglas-fir nursery seedlings (modified from Ritchie 1984a).*

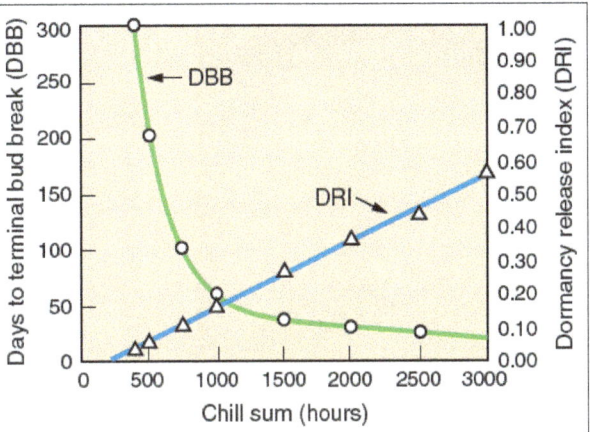

Figure 7.2.34—*Because days to bud break (DBB) over chill sum is a curvilinear relationship, it is useful to convert it to a linear dormancy release index (DRI). In this example, DRI = 10/DBB because Douglas-fir seedlings resumed growth (broke bud) in 10 days when their full chilling requirement was satisfied (modified from Ritchie 1984a).*

(Sorensen 1983, Lindqvist 2000). After this curve has been developed for a nursery, it can be used subsequently to estimate dormancy intensity for a given species and seed zone directly from chilling sums.

From this experiment, it is clear that bud dormancy intensity is very high in fall and drops sharply in early winter, in contrast to the common misconception that deepest dormancy occurs in midwinter when plants are most stress resistent. In addition, this test illustrates that there is no simple "chilling requirement" for any species. Rather, there is a curvilinear relationship between chilling and dormancy in which more chilling will result in more rapid budbreak under forcing conditions. For example, Douglas-fir seedlings with only 800 hours of chilling exposure will eventually break bud, but not nearly as rapidly as those exposed to 2,000 hours of chilling (fig. 7.2.33).

Calculating the dormancy release index. Now that DBB for a given crop can be estimated from chill sums, how is this information used? If DBB were measured on a group of Douglas-fir seedlings that were fully released from dormancy (that is, the chilling requirement was completely fulfilled), the buds would break in about 10 days. Taking this number as the denominator, an index can be calculated that expresses the dormancy intensity on a linear scale:

$$\text{Dormancy release index (DRI)} = 10/\text{DBB}$$

DBB is the days to bud break of a test group of plants as described in the experiment above.

Buds at peak dormancy have a DRI value near zero (for example, DRI = 10/300 = 0.03). As dormancy weakens, DRI approaches 1 (for example, DRI = 10/15 = 0.67). This relationship is shown in figure 7.2.34. DRI is useful because it transforms the curvilinear relationship between dormancy intensity and chilling sum to a more useable linear form. This linear regression can then be used to provide a benchmark and common scale for comparing stock lots in a given plant species.

McKay and Milner (2000) developed a variation on this approach; they estimated DRI by counting the days required for 50 percent of the terminal buds to break in Sitka spruce, Douglas-fir, Japanese larch, and Scots pine. Their results also closely resemble those of figure 7.2.34. The DRI has been particularly useful as an indicator of plant stress resistance—a key performance attribute. We will discuss this relationship and how it is used in Section 7.2.5.2.

 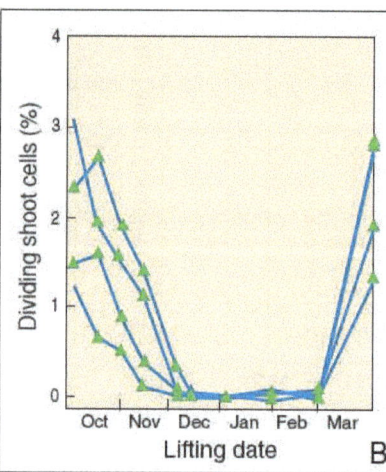 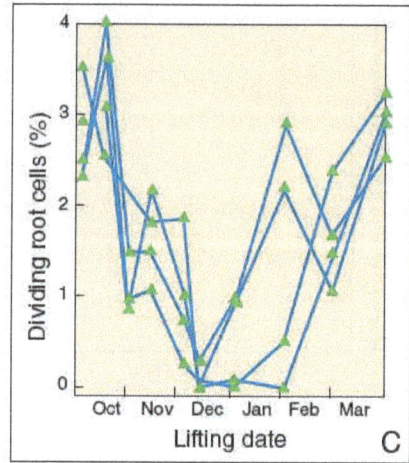

Figure 7.2.35 —*Measuring cell division rates in buds (A) is a laboratory measure of dormancy. Shoot activity over 4 years shows a characteristic pattern of inactivity during winter (B), but roots (C) continue to grow whenever conditions are favorable (modified from O'Reilly and others 1999).*

Measuring mitotic index. In our definition of dormancy, we stressed that dormancy referred only to buds or other plant meristems (fig. 7.2.29). Laboratory techniques have been developed to measure the number of meristematic cells that are dividing at any given time (fig. 7.2.35A). Although primarily used for research purposes, these measurements also illustrate dormancy patterns.

For example, the tips of terminal shoots and long roots of bareroot Douglas-fir seedlings were excised and, after examining meristematic cells with a 400X microscope, a mitotic index was calculated (O'Reilly and others 1999). The results indicate that terminal bud activity shows a definite seasonal pattern; cell division slows gradually in fall and stops completely during winter. With warmer temperatures and longer days in late winter and early spring, cell division begins to increase rapidly (fig. 7.2.35B). This is in direct contrast to the patterns of root meristem activity, showing that roots never become truly dormant but will grow whenever soil temperatures permit (fig. 7.2.35C). Although useful to researchers, this test is too time consuming to be used operationally.

Bud size and development. Although bud size and development are not, in themselves, indicative of the intensity of bud dormancy, they have traditionally been viewed by nursery managers as an indicator of plant quality. For example, a bud length measuring protocol was developed by the Ontario Ministry of Natural Resources as part of their former quality testing service. The process involves cutting buds in half and counting needle primordia. At the end of the hardening phase, low numbers of primordia were interpreted to indicate stressful conditions and increased susceptibility of overwinter damage. Conversely, seedlots having buds with large numbers of needle primordia were rated as being of higher quality (Colombo and others 2001).

Dormancy: Summary. Although the term "dormant plants" is common in nursery jargon, dormancy refers only to meristematic tissues of the shoot: buds and lateral cambium. Bud dormancy has been studied most intensively and is of major interest to plant producers and users.

Forest and conservation nursery crops, like all perennial plants, undergo an annual cycle of activity. In late summer, shortening photoperiods trigger plants to begin the bud dormancy process that culminates in late fall. This condition is known as deep dormancy and can be released by exposure to a period of low temperatures. This process is known as satisfying the chilling requirement, and temperatures in the range of about 3 to 5 °C (37 to 41 °F) are most efficient. By late winter, the chilling requirement has been met and buds will break whenever temperatures permit.

Unfortunately, bud dormancy cannot be quickly or easily measured. The only reliable method is to conduct a bud break test by bringing samples of plants into a forcing greenhouse at regular intervals throughout winter and recording the days required for the buds to break (DBB). After the relationship between DBB and chilling has been developed for a nursery, it can be used to establish harvesting windows and to estimate the dormancy intensity of crops during subsequent winters.

A useful index of dormancy intensity, the dormancy release index, makes the DBB information more practical by converting the data to a straight line.

Although we lack a rapid test for bud dormancy, it can be estimated from the known relationship between chilling and dormancy intensity as measured by DBB. Nurseries can measure the chilling requirement for their various crops and use this information to monitor the release of bud dormancy.

7.2.5.2 Stress resistance

In the previous section, we indicated that dormancy is closely related to stress resistance (SR). From an operational standpoint, we will introduce some techniques that nursery managers can use to estimate the relative SR of a crop at any point during the harvesting-to-outplanting process.

The concept of stress resistance. Plants are subjected to a variety of stresses (mechanical stresses, root exposure, rough handling, and desiccation, to name just a few) from the time they are harvested in the nursery to when they are outplanted. Nursery managers use a variety of cultural techniques, collectively termed "hardening-off," to prepare their stock to tolerate these stresses. Realizing its importance and practical applications, plant physiologists have been studying SR for almost 40 years.

Hermann (1967) determined that SR was related to root system function in bareroot stock, and Lavender (1984) showed that SR varies seasonally, reaching a midwinter peak after bud dormancy intensity has begun to decline (fig. 7.2.36). The data for this seasonal curve came mainly from outplanting trials, which is why it corresponds exactly with the traditional midwinter lifting season.

Obviously, nursery managers want to maximize SR in their crops and maintain this condition until they are shipped to

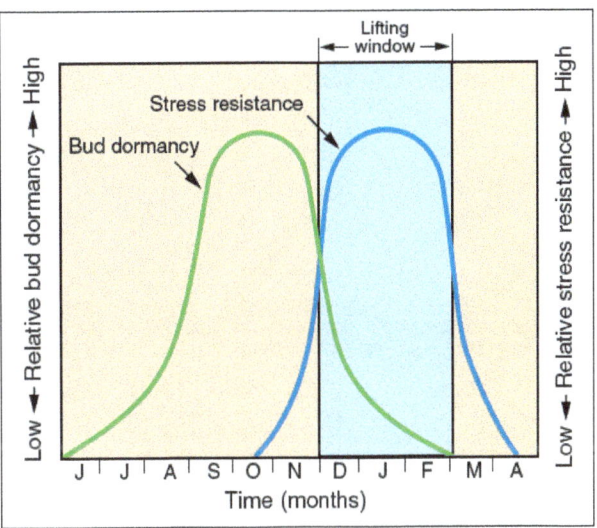

Figure 7.2.36—*This classic illustration shows that bud dormancy and stress resistance follow similar bell-shaped trajectories but occur at different times. Comparison to the traditional midwinter lifting window shows that stress resistance is a better indicator of when to harvest (lifting window) and store nursery crops (modified from Lavender 1984).*

their customers for outplanting or transplanted back into the nursery. But how can they measure or estimate SR, and how can they culture their crops to achieve maximum SR?

Measuring stress resistance. A quick and easy way to measure the SR of nursery stock would be an invaluable tool, and there have been many attempts to develop a test to ascertain this important aspect of quality.

Stress tests. During the 1970s and 1980s, several attempts were made to develop quick tests of SR. For example, a stress test was developed at Oregon State University (McCreary and Duryea 1984) that consisted of harvesting plants, potting them, and exposing them to stressful conditions, mainly high temperature, low humidity, and low soil moisture. After a predetermined time, plants would then be moved into a greenhouse and, after several weeks, be assessed for survival, root growth, bud break, and other indicators of vigor (fig. 7.2.37). Despite some promising early results, the outcomes of literally hundreds of such tests proved difficult to interpret and not very repeatable. Accordingly, this quality test was abandoned.

Figure 7.2.37—*Stress tests involve harvesting seedlings and exposing them to a stressful environment. At Oregon State University, the stress was a dry, hot greenhouse.*

Another more elaborate and more time-consuming, but more accurate, method of measuring SR involves a procedure similar to cold hardiness testing (Ritchie 1986). It consists of three sequential steps:

1. Exposing plants to a controlled stress treatment. The most commonly used stress treatments employ some sort of controlled trauma to root systems. This might involve exposure to high or low temperatures, prolonged drying, or a simulation of rough handling, such as dropping or tumbling.

2. Outplanting stress-treated plants into a natural environment where their growth response to the treatment can be expressed. By "natural," we mean the plants should be growing in soil and exposed to the ambient outdoor environment, but they must be able to express growth potential without confounding effects of browsing, water stress, or weed competition. A bareroot nursery bed that is watered regularly and kept weed free is ideal. The test plants are set out in replicated blocks along with nonstressed controls of similar initial size from the same seedlots or families.

3. Evaluating the impact of the stress treatment by comparing the performance of the stressed plants to that of nonstressed controls after a predefined time period, typically one complete growing season. The assessment can be as simple as measuring shoot growth or as complicated as destructively sampling the entire plant and measuring total biomass. We have found that removing the shoot of the plant and determining its dry weight is a good basis for comparison. In this approach, SR is characterized as the difference in growth between the stressed plants and nonstressed controls. A helpful way of expressing this difference numerically is by calculating a stress injury index (SII) using the first-year shoot growth of the stressed (G_s) and nonstressed control seedlings (G_c):

$$SII = 100 - (G_s/G_c \times 100)$$

The SII expresses the percentage reduction in top growth resulting from stress injury, and so, the lower the value, the higher the stress resistance of the test plants (Ritchie and others 1985).

Using cold hardiness tests to estimate overall stress resistance. Decades of nursery experience have shown that, when plants are at their maximum state of hardiness, they are the most resistant to the many stresses of harvesting, handling, storage, shipping, and outplanting. In fact, recent genetic research has revealed that some of the same (dehydrin) gene complexes that are involved in cold acclimation also play a key role in resistance to water stress (Wheeler and others 2005).

Container nurseries in western Canada use a "storability test" to determine if plants are physiologically ready for harvesting, packaging, and cold storage (Simpson 1990). Essentially, if plants are cold hardy to a threshold temperature of $-18\ °C$ ($0\ °F$), then they are ready to withstand the stresses of storage. A more recent modification that uses chlorophyll fluorescence (see Section 7.2.4.4) to determine

Table 7.2.10—*Seedling quality classes based on dormancy release index (DRI) and stress resistance (SR) (modified from Ritchie 1989)*

Quality class	DRI value	Degree of SR
Class 2	< 0.25	Seedlings are below peak SR but are increasing.
Class 1	0.26 to 0.40	Seedlings are at peak SR.
Class 3	> 0.40	Seedlings are beyond peak and SR is decreasing.

if tissue damage has occurred and produces results up to 6 days earlier than visual evaluation (L'Hirondelle and others 2007). Because this method tests plant samples directly, it has proved to be a reliable predictor of outplanting performance (Kooistra 2003). A similar storability test based on FIEL is used in container nurseries in Ontario (Colombo 2009). To use this test in a more temperate or coastal area, a higher temperature threshold would need to be determined.

Using chilling hours to predict stress resistance. It is intuitive that SR is very closely related to dormancy, and this has been verified by plant physiology research (Ritchie 1986, 1989; Ritchie and others 1985). As dormancy intensity weakens through winter in response to chilling, SR gradually increases to a midwinter high. Then it falls rapidly as dormancy is fully released and spring approaches (fig. 7.2.38). The physiological mechanisms behind this relationship are not fully understood, but it is repeatable from year to year with different crop types (bareroot and container) and species (Douglas-fir, pines, spruces, some hardwoods) and across nurseries (Burr and others 1989; Cannell and others 1990; Ritchie and others 1985). This means that if you can track the dormancy status of a crop through winter, this information can be used to estimate SR without measuring it directly.

As discussed in the previous section, bud dormancy peaks in fall and is released gradually during winter as plants are exposed to low temperatures—the "chilling

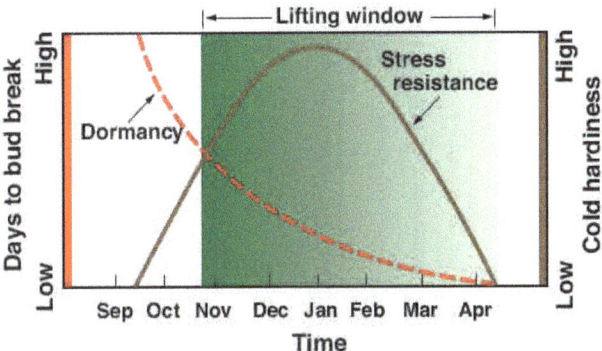

Figure 7.2.38—*Both bud dormancy, as measured as days to bud break (DBB), and stress resistance, as measured by cold hardiness tests, can be used to determine the best time to harvest nursery stock (lifting window). Cold hardiness tests, however, are so much quicker and easier that they have become the standard test for lifting and subsequent refrigerated storage.*

requirement." Transforming this curvilinear relationship into a linear dormancy release index (DRI) makes it much easier to use. The DRI is 0 at peak dormancy in fall, and approaches 1 as dormancy is released in spring.

Research with Douglas-fir has revealed a consistent relationship between DRI and SR (Ritchie 1986). In early winter, when DRI is in the range between 0 and about 0.25, SR is low but increasing. Between DRI 0.26 and 0.40 (midwinter), SR reaches a seasonal high, but when DRI exceeds 0.40 (early spring), SR declines and plants become very susceptible to damage. These results lead to the definition of three seedling quality classes based on dormancy intensity and SR (table 7.2.10).

After the relationship between chilling and DRI has been established for a given species in a given nursery, it can be used to estimate SR at any point during the winter for subsequent crops at that nursery. Let us say, for example, that it is late December and your nursery chilling sum is about 1,000 hours. Using figure 7.2.39, you would estimate that DRI was approaching 0.2. From table 7.2.10, we see that stock at this time is in SR Class 2—not yet peaked, but will improve with more chilling. Now, let's say it is February and you have about 2,000 hours of chilling at your nursery. DRI is about 0.38, indicating that SR is in the seasonal high range but will soon begin to decline.

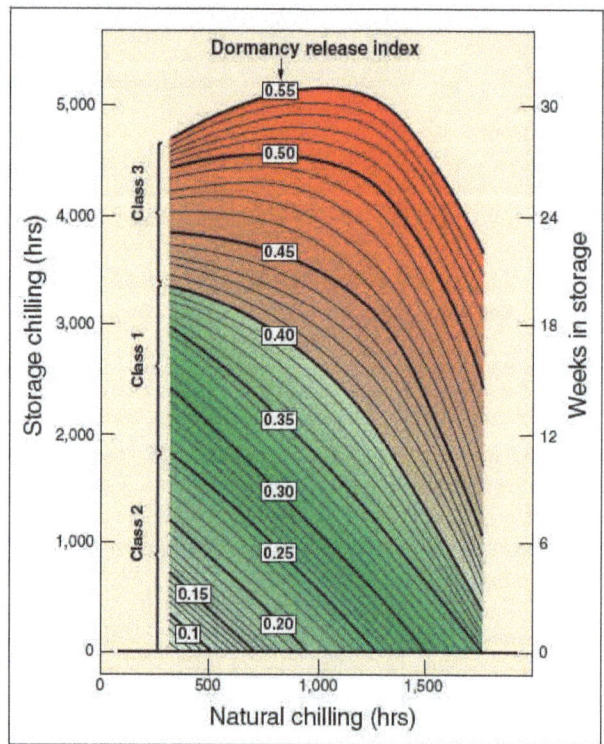

Figure 7.2.39—*Graph showing how the chill sum at time of lifting, combined with time in cooler or freezer storage, can be used to predict the dormancy release index (DRI) and stress resistance class (table 7.2.10), of planting stock. The graph is entered on the X-axis at the nursery chill sum at which seedlings were placed into storage. The storage duration is found on the Y-axis. These lines intersect at the DRI value of the seedlings at that time. Their quality class can then be read from the Y-axis (modified from Ritchie 1989).*

Adjusting for the added effect of refrigerated storage. For crops transplanted or outplanted without cooler or freezer storage ("hot-planted"), DRI is very useful. You simply look at the chilling sum at any point and, from it, estimate stress resistance. But many nursery crops are refrigerated from a few weeks to several months before transplanting or outplanting. So, how does that affect SR?

The low temperatures in refrigerated storage are within the chilling range; hence, they contribute to dormancy release. They do so inefficiently, however, because storage temperatures are below the optimum chilling temperature (Ritchie 1984a, van den Driessche 1977). Therefore, *refrigerated storage has the effect of slowing the release of dormancy.* This means that plants harvested and placed into refrigerated storage will pass through SR Classes 2, 1, and 3 more slowly than they would if left in open container storage (see Chapter 7.4). Plants that are kept in freezer storage accumulate very little chilling because temperatures are well below optimum. These plants must have already accumulated an adequate level of chilling prior to being placed in storage.

To use the graph, select total ambient chilling hours from a nursery on the X-axis. For this example, let us use 1,000 hours. At this point, the stock will have a DRI value of about 0.20, placing it in Quality Class 2 (table 7.2.10). Now, if the plants are held in refrigerated storage for about 4 weeks, they will enter Class 1 and have even higher SR. However, if these same plants had been held in the nursery for a few more weeks until they accumulated over 1,300 hours of chilling, they would exceed the DRI limit of 0.25 and enter Class 1 and have maximum SR. Then, if they were placed in freezer storage, they could be held for at least 15 weeks (right axis) before their DRI approached 0.40 and their quality dropped to Class 3. (Note: as a rule of thumb, cooler storage should not exceed 6 weeks. If storage longer than 6 weeks is needed freezer storage should be used—see Chapter 7.3.)

On a practical basis, figure 7.2.39 integrates the effect of both harvesting date and storage duration on DRI and, hence, stress resistance. If the chill sum at the time of harvesting is known, then storage duration can be planned to deliver stock when it is at maximum SR: Class 1. If the planned outplanting date is known, then lift date and time in storage can be prearranged to deliver stock to the outplanting site so it will be in Class 1. This graph illustrates the very important point that, for outplanting sites that cannot be accessed until late, early winter lifting with overwinter freezer storage is preferable to late spring lifting with or without storage.

Application to other species and regions. The data that were used to produce figure 7.2.39 came from coastal Douglas-fir seedlings from four different seedlots (high and low elevation lots in both Washington and Oregon) that were grown in two different coastal nurseries (Washington and Oregon). These results have been operationally tested with Douglas-fir crops from other seedlots and during other growing seasons with consistent results. Therefore, for West Coast nurseries raising Douglas-fir, figure 7.2.39 is a very handy way of estimating SR from chilling hours.

For interior or northern nurseries, however, the relationship between chilling and DRI may be quite different. This was tested in an interior west Canadian nursery with lodgepole pine and interior spruce (Ritchie and others 1985). The results showed that chilling began to accumulate earlier in fall and that more chilling accumulated throughout winter. The results also suggested that these species may require more chilling hours for full dormancy release than coastal Douglas-fir, similar to results with ponderosa pine (*Pinus ponderosa*) (Wenny and others 2002). Nevertheless, the overall relationships (if not the same numbers) shown in figure 7.2.39 were similar to what has been found with Douglas-fir. Therefore, to accurately predict SR from chilling hours for other species and nurseries, a chilling-DRI "calibration curve" needs to be developed.

Stress resistance: Summary. Stress resistance (SR) is an important, but elusive, performance attribute that describes a plant's ability to tolerate the stresses associated with harvesting, handling, storing, and outplanting. SR varies seasonally; it is low in fall, high in midwinter, and low in spring.

SR is very laborious to measure, so no operational test is currently being used. However, because the seasonal pattern of SR closely coincides with the pattern of cold hardiness, standard cold hardiness tests can provide quick and useful estimates of SR.

Studies have shown that SR is related to dormancy intensity expressed as a dormancy release index (DRI). When DRI is in a range between 0 and about 0.25, SR is low but improving. Between DRI 0.25 and 0.40, SR is at a seasonal high. Above DRI 0.40, SR is declining. Most important, this relationship tends to be consistent whether or not plants have been stored.

Because cooler and freezer storage slows the release of dormancy, storage prolongs the period of high SR. These relationships can be used to schedule harvesting and storage in order to deliver stock to the planting site that has very high resistance to stress. Although most of this research was done with bareroot stock of commercial conifers, the basic principles should apply to container plants of other species.

7.2.5.3 Root growth potential

Although Wakeley (1954) published the first account of the relationship between new root growth and plant quality, it was Stone (1955) who, after experimentation, coined the term "root regenerating potential" to describe his new indicator of seedling physiological quality.

Basing their effort on Stone's original research, other workers began developing and using this method of plant assessment (for example, Burdett 1979; Jenkinson 1975). A comprehensive review of root growth potential (RGP) by Ritchie and Dunlap (1980) was responsible for a flurry of new research and adoption of RGP as the first performance quality test used operationally in forest nurseries. Because of this wide interest, a chapter on Assessing Seedling Quality in the *Forest Nursery Manual* (Duryea and Landis 1984) featured a discussion and strong endorsement of RGP (Ritchie 1984b). Further reviews (Duryea 1985; Ritchie 1985; Ritchie and Tanaka 1990) made this test the most popular and widely used quality test (fig. 7.2.40A). RGP tests have been employed worldwide and have been the subject of much discussion (Binder and others 1988; Landis and Skagel 1988; Sutton 1983) and even debate (Simpson and Ritchie 1997).

RGP test procedure. The RGP test consists of placing a random sample of plants into an environment that promotes rapid root growth. After 7 to 28 days, the plants are evaluated for new root growth. In the following section, we examine each step in the process.

Sampling. As with all tests, if sampling is biased (not random), test results will be meaningless. The number of plants used in a typical RGP test is quite small and should be randomly selected from the population at large in order to be as representative as possible. A sample of 60 seedlings, which is the number usually required by testing laboratories, is only 0.12 percent of a moderately sized seedlot of 50,000 seedlings. A 25- to 30-plant sample would be a minimum number to evaluate.

It is simple in principle to collect a random sample when plants are still in containers or on the grading table, but sampling becomes more difficult after stock has been packaged and stored. When cooler stored, it is operationally difficult to sample from bagged plants, because a number of bags must be accessed, opened, and the sample collected from throughout the bag, not just from the top layer of plants. Sampling during freezer storage requires special packaging (Landis and Skagel 1988).

Time of sample collection. Tests performed on plants at the time of harvesting are useful to evaluate nursery cultural practices but may not reflect the condition of the plant at time of outplanting. If you are interested in outplanting performance, then the best time to sample is as close to the time of outplanting as operationally possible (Landis and Skagel 1988).

Test environment. The testing environment is particularly important because it must provide conditions that are near "optimum" for root growth (Landis and Skagel 1988). The temperature should be 19 to 25 °C (66 to 77 °F). The rooting medium should be well aerated and watered, and there should be adequate light and long days. Because these factors will affect test results, it is important to maintain consistent conditions across tests, although this can be difficult.

Three types of test environments have been used:

Pots in greenhouse—Most quality-testing facilities use this method, in which plants are potted in 3.8 to 7.6 liter (1 to 2 gal) containers filled with a well-drained artificial growing medium. The pots are kept well irrigated in a greenhouse (fig. 7.2.40B) for the duration of the testing period (Ritchie 1985; Tanaka and others 1997). After 7 to 28 days, the growing medium is washed from the roots (fig. 7.2.40C) and the amount of new root growth is rated.

Hydroponic—Plants are suspended with their roots in warm, aerated water, such as in an aquarium. This method has found use with several deciduous hardwood species (Wilson and Jacobs 2006).

Aeroponic—Plants are suspended in a closed chamber while warm water mists the roots (fig. 7.2.40D). Forest Service nurseries have used this technique with good results (Rietveld and Tinus 1990). One benefit is that the rack of plants can be easily removed from the misting chamber to monitor root development during the test period (fig. 7.2.40E).

Evaluation. After the test is completed, new root growth must be quantified. Researchers have attempted to shortcut this tedious process using photography, dyes, root volume measurements, and other approaches. Despite this, the tried-and-true "root count" technique has prevailed. This involves visually estimating the number of new roots greater than 1 cm (0.4 in) long on the plant. An experienced technician can do this in a few minutes. This count can be reported as a raw number (for example, 120 roots per plant) or transformed into an index such as reported by Burdett (1979) and modified by Tanaka and others (1997) (table 7.2.11). Root numbers and total root length are usually well correlated.

RGP as a predictor of ouplanting performance. Interpretation of the results of RGP tests remains challenging. A common misconception has been to assume that RGP results directly predict outplanting performance. In other words, high RGP always ensures high survival, while low RGP always ensures low survival (fig. 7.2.41A). At best, RGP is positively correlated with survival only about 75 percent of the time (Ritchie and Dunlap 1980, Ritchie and Tanaka 1990). Sometimes these correlations are weak, sometimes strong. Binder and others (1988) found no correlation between RGP and outplanting mortality in 8,600 operational trials in British Columbia. This is because the outplanting environment (which is usually very different from the RGP testing environment) has an overriding influence on performance (Binder and others 1988; Landis and Skagel 1988; Simpson and Ritchie 1997; Sutton 1983). Performance of low-RGP stock on harsh sites and of high-RGP stock on mild sites is usually predictable. However, performance of low-RGP stock on mild sites and high-RGP stock on harsh sites is not (fig. 7.2.41B).

Figure 7.2.40—*Because the relationship between new roots and outplanting success is intuitively important (A), the root growth potential test quickly became the most popular and widely used assessment of plant quality. One testing procedure involves growing test plants in pots in a greenhouse (B), washing roots (C), and then rating the amount of new root growth. In the second procedure, test plants are supported in a mist chamber (D) and then measured for the length and number of new roots (E).*

Table 7.2.11—*Root growth index (RGI) scale developed by Tanaka and others (1997) to quantify root growth following a root growth potential (RGP) test*

Root growth index (RGI)	Number of new roots 1 cm or longer
0	None
1	Some roots but none > 1 cm
2	1–3
3	4–10
4	11–30
5	31–100
6	101–300
7	More than 300

It seems intuitive that for a newly outplanted plant to survive and grow, it must rapidly regenerate new roots in order to maintain an adequate water balance. This logic has been used to explain why RGP can be expected to predict survival. Simpson and Ritchie (1997), however, point out that newly planted stock is almost never able to grow roots after outplanting because, although soil moisture may be high, soil temperature during the winter or early spring planting season in most places is far below the threshold temperature for root growth (fig. 7.2.41C). Under these conditions, the existing root system is adequate to supply water to the plant until the soil warms and roots begin to grow (McKay 1998). Therefore, whether or not new root growth occurs immediately after planting is of little consequence to field performance.

Why RGP often works. The discovery that many conifer seedlings, especially Douglas-fir, require mainly current photosynthate for new root growth (van den Driessche 1987, 1991) has provided a rationale for interpreting RGP test results. For a plant to grow new roots in the test environment, the foliage must be photosynthesizing (fig. 7.2.42). Therefore, the stomata must be open, the leaves must be healthy, and the photosynthetic apparatus must be functioning properly. Photosynthate must move to the root system, so the phloem pathway to the roots must be intact, and the roots themselves must be metabolizing normally. If any of these systems have been compromised by, say, cold damage, water stress, disease, photodamage, or other agents, a depression of RGP will result.

Taken in that light, then, a more realistic view is that RGP testing is analogous to seed testing, which provides a snapshot of seed viability at the time seeds are tested. No one would expect seeds that had 95 percent laboratory germination to always give 95 percent emergence in the nursery. But if the test gave an abnormally low value, it would indicate poor seed viability. This is the model to use when interpreting RGP test results. The RGP test is a "red flag" test that identifies stock lots that, for whatever reason, are not up to par.

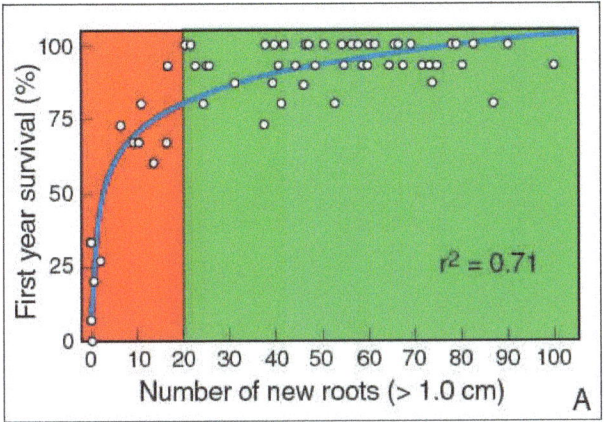

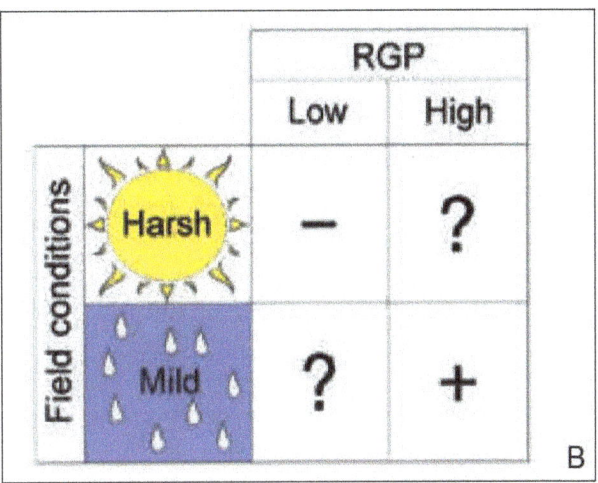

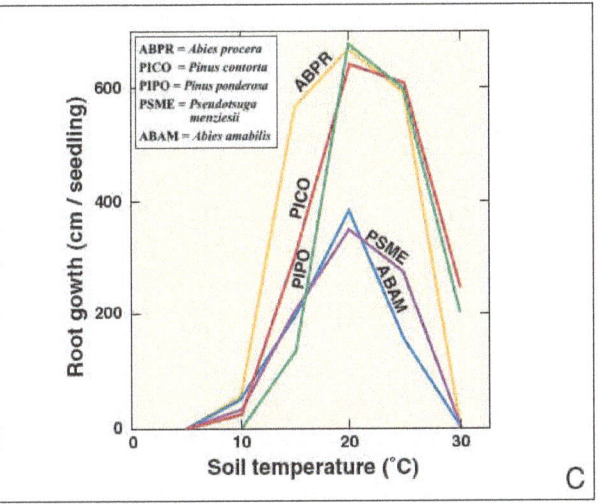

Figure 7.2.41—*Although a good relationship between root growth potential (RGP) test values and outplanting success sometimes exists (A), limiting factors on the outplanting site often prevents good predictability. Performance of low-RGP stock planted on a harsh site or high-RGP stock on a mild site is generally predictable. However, performance of high-RGP stock on a harsh site, or low-RGP stock on a mild site is not (B). One frequent problem is that soil temperatures on the outplanting sites are much lower than the ideal temperatures used in the testing environments (C) (A, modified from Grossnickle 2000; C, modified from Lopushinsky and Max 1990).*

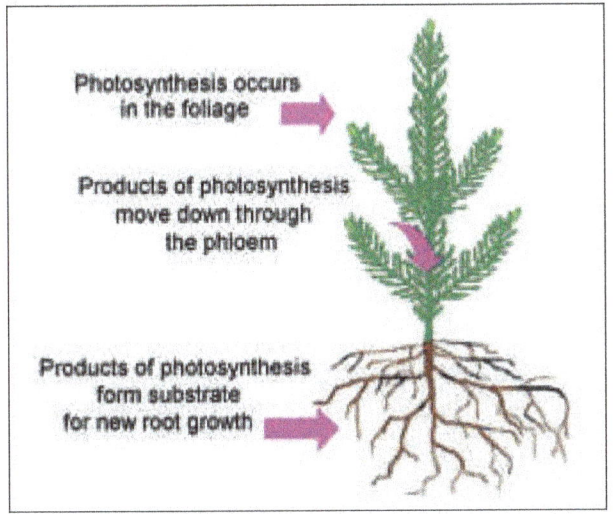

Figure 7.2.42—*Root growth in many conifers depends on a supply of current photosynthate from the shoot (van den Driessche 1987, 1991). Any factors that depress photosynthesis or impede the flow of photosynthate from leaves to roots will result in reduced root growth potential.*

Root growth potential: Summary. RGP remains the most popular quality test because it is intuitive, robust, and simple. Like any test, however, RGP has its limitations. The major drawback of the RGP test is the long testing period and the limited predictive ability. RGP tests provide only a "snapshot in time," because plant physiological quality can change right up until the stock is outplanted.

RGP sometimes predicts survival and other times does not. This is because site conditions, which are very different from the testing conditions, can override stock quality. RGP does not predict root growth after outplanting, and root growth after outplanting generally has little to do with survival.

The RGP test is a valuable test of viability—that is, it determines whether plants are alive and functional at the time the test is conducted. RGP test results integrate many physiological systems in plants, such as stomatal function, the photosynthetic mechanism, phloem integrity, root viability, seedling nutrition, and so on. If any of these systems have been compromised, it will show up as a depression of RGP.

Regardless of their predictive value, RGP tests have been done long enough to show that nursery stock with high RGP values will have great survival and growth (Maki and Colombo 2001). Results of an RGP test should be interpreted in the same way as results of a seed germination test. It is a "red flag" test that identifies sub-par lots and may or may not predict field performance.

7.2.6 Correlating Combinations of Plant Quality Tests To Predict Outplanting Performance

As you should have deduced by now, nursery plant quality is a complicated subject. So, instead of trying to predict outplanting performance with just one variable, it makes sense to attempt correlations with two or more plant quality indices. Research to develop a comprehensive approach that uses a battery of tests has been done (Grossnickle and others 1991) but has not been adopted operationally. Recent research in British Columbia measured root growth potential, chlorophyll fluorescence, and stomatal conductance of conifer seedlings and then correlated them singly and in combination with survival and growth after outplanting (L'Hirondelle and others 2007). They found that, while survival was highly correlated with root growth potential ($R^2 = 0.72$), the combination of root growth potential and chlorophyll fluorescence was a good predictor of survival plus shoot growth as measured by dry weight (fig. 7.2.43). We hope more research will be done in this area to further refine our ability to mathematically predict nursery stock quality.

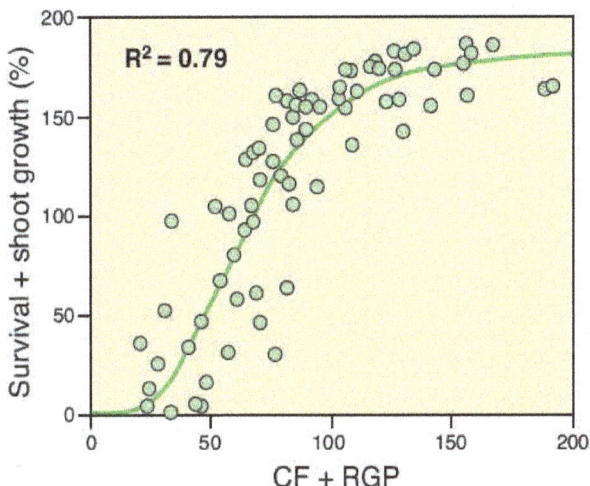

Figure 7.2.43—*Measuring root growth potential and chlorophyll fluorescence proved to be a good predictor of total outplanting performance (survival + shoot growth) of conifer seedlings (modified from L'Hirondelle and others 2007).*

7.2.7 Limitations of Plant Quality Tests

7.2.7.1 Timing

Each plant quality test we have discussed should be done at a particular time in the nursery-through-outplanting cycle. Morphological attributes change as the crops grow in the nursery but remain constant after harvesting. Physiological and performance attributes, however, vary considerably depending on when the measurements are taken. For instance, plant moisture stress has a pronounced diurnal pattern whereas cold hardiness increases during the fall and can be lost during refrigerated storage (Sundheim and Kohmann 2001). Root electrolyte leakage and chlorophyll fluorescence (CF) are used mainly to detect damage following a stress event. Therefore, they should be measured immediately after the event, while keeping two important considerations in mind. First, to know whether test results are "normal," baseline information on these variables must be available. That often calls for routine monitoring of these variables in healthy crops prior to the stress event. The second, and very important, point is that plants may require time to exhibit stress symptoms and also have the ability to recover from stress. So, for example, CF values measured the day after a cold event may not give an accurate picture of the damage sustained by the crop or of its longer-term response.

Both nursery managers and seedling users can use plant quality tests but would do so at different times. For example, a nursery manager would use plant moisture stress to schedule irrigation and cold hardiness tests to determine lifting windows and storability, whereas a seedling user might use plant moisture stress to ensure that nursery stock was not moisture-stressed prior to outplanting and cold hardiness tests to indicate overall stress resistance before outplanting (fig. 7.2.44).

7.2.7.2 Sampling

Proper sampling is critical to effective seedling quality testing. If the sample is biased, the test results will be biased and therefore worthless. One wonders how many of the quality tests that failed to predict field performance were conducted on samples that did not adequately represent the populations from which they were drawn. It is important to follow the "three Rs" of sampling: random, replicated, and representative. Multiple samples collected randomly from throughout a given crop will yield the

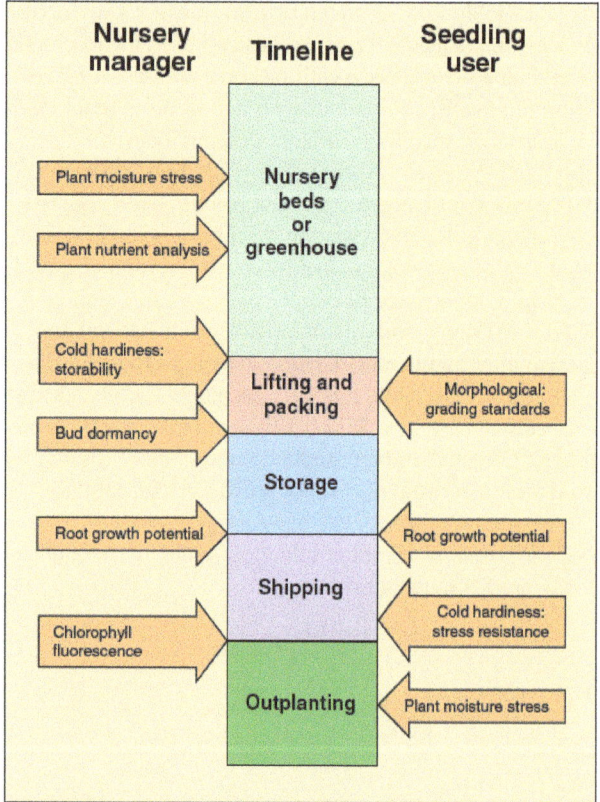

Figure 7.2.44—*Plant quality testing can be done by both nursery managers and seedling users. The timing of the various tests will vary with the desired interpretation.*

most useful data. Many growers are resistant to spending the time or money to collect and test samples in this manner. If you think about it, however, spending a relatively small amount of time and money on a single, biased test is simply wasted time and money to generate meaningless data, whereas spending a little more time and money using a three R sampling protocol produces valuable data which can assist with management decisions.

7.2.7.3 Unreasonable expectations

It is important that plant producers and users employ the right test at the right time and that they remain aware of the pitfalls of reading too much into test results. A discussion of this topic can be found in Simpson and Ritchie (1997) who propose the following conceptual model of field performance:

$$\text{Field performance} = f(SC, PM, SR, PV)$$

where:

SC = site conditions (all physical, chemical, and biological characteristics of the site during and after planting),

PM = plant morphological attributes (stem diameter and height, shoot-to-root ratio, root quality, and so on),

SR = stress resistance (ability to withstand stresses associated with harvesting, storage, handling, and planting), and

PV = plant viability (freedom from disease, injury, or stress-induced disorders); plant "functional integrity" (Grossnickle and Folk 1993) is a good way to express this idea.

Obviously, quality testing does not provide information on SC, but it can yield detailed information on PM and can offer insights on SR through monitoring of cold hardiness and dormancy intensity. PV can also be approximated using root growth potential, chlorophyll fluorescence, root electrolyte leakage, and, to some extent, plant moisture stress.

With this "package" of available quality tests and protocols, nursery managers have sufficient tools to make more than an educated guess about the quality of any given stock lot at any given time. But, it should be remembered that quality must be viewed within the context of site conditions that can never be fully predicted.

7.2.8 Commercial Plant Quality Testing Laboratories

Several of the tests enumerated above can be administered on the nursery site (for example, root electrolyte leakage, root growth potential, chill sum accumulation). Certain tests (for example, cold hardiness and chlorophyll fluorescence), however, require elaborate and expensive equipment. Seedling quality laboratories typically use equipment such as growth chambers, which generate more uniform, replicable test conditions. Using a testing service has the added benefit of providing an independent assessment of seedling quality. Over time, these assessments can be organized in a database to reveal patterns that might not otherwise be apparent (Colombo 2009).

At the time of this writing (2009), we are aware of four laboratories in North America that provide quality testing services. They are listed in appendix 7.2.1.

7.2.9 Summary and Conclusions

Plant quality is divided into three broad categories of attributes: morphological, physiological, and performance. Morphological attributes are easy to see and measure and do not change readily after plants are harvested and stored. Container size and plant density have the most pronounced effects on morphology. Although many characteristics may be measured (for example, shoot height, stem diameter, biomass) and ratios of those characteristics can be calculated (for example shoot-to-root ratio), shoot height and stem diameter are the most frequently measured morphological traits and the most commonly used grading criteria. Initial shoot height tends to be correlated with height growth after outplanting, whereas initial stem diameter is better correlated with survival.

Physiological attributes are not readily visible and require specialized equipment and testing to ascertain. Evaluations of plant moisture stress, cold hardiness, root electrolyte leakage, and chlorophyll fluorescence are most common.

Plants lose water more rapidly through transpiration than they can absorb from the soil, putting the plants under "plant moisture stress" (PMS). This level of stress can be quantified by using a pressure chamber. Although a direct correlation between PMS and any of the classical plant quality indicators is lacking, nursery managers can use pre-dawn PMS measurements to schedule irrigation and to monitor stress during hardening, harvesting, and outplanting.

Development of cold hardiness in nursery stock is triggered by changes in photoperiod in late summer and increases rapidly in late fall and early winter as plants experience lower temperatures. For temperate zone plants, peak hardiness occurs in January and is quickly lost as photoperiods lengthen and temperatures increase. Different plant parts may have different cold hardiness levels; buds are generally most cold hardy while roots are the least. Cold hardiness levels can be determined using a whole plant freeze test, freeze-induced electrolyte leakage (FIEL), or analysis of genetic indicators. Results from testing can be used by nursery managers to decide safe windows for harvesting, to provide necessary frost protection, and as a surrogate for estimating overall stress resistance.

Assessing root electrolyte leakage (REL) is similar to FIEL, but it is broader because this test looks at potential loss of root viability from many factors, such as disease, rough handling, and desiccation, and not just at damage from cold temperatures. It is difficult to correlate REL with plant survival because many factors other than root damage can affect REL.

Chlorophyll fluorescence provides a means of determining a plant's ability to efficiently photosynthesize. Stresses, whether they are short term, subtle, long term, or severe, can impair this important physiological process. This measurement can identify when significant damage to the photosynthetic system has occurred, indicating a plant's performance may be compromised. More work is needed before this test will be an operational quality test.

Performance attributes integrate both morphological and physiological attributes. Although testing performance attributes has great value, these tests can be laborious and expensive. Measures of dormancy, stress resistance, and root growth potential (RGP) are the common tests.

Although nursery managers talk about dormant plants, dormancy only refers to meristematic tissues, and only bud dormancy has been extensively studied. Shoots may cease to elongate and form buds in response to changing environmental conditions that are less favorable to growth (quiescence), or in response to reduced photoperiod (deep dormancy) that culminates in fall. Once deeply dormant, buds require a specific duration of exposure to cold temperatures (chilling) before shoot growth will resume. The chilling requirement is the length of exposure to cold temperatures that buds need before they are once again quiescent and ready to resume growth when temperatures permit. The only reliable way to estimate the intensity of bud dormancy is to measure how much chilling buds have been exposed to, and then record how many days are required for those buds to resume shoot growth (days to bud break—DBB) when they are returned to favorable growing conditions. The relationship between chilling and DBB is curvilinear, but a simple dormancy release index (DRI) can be used to convert the data to a straight line and make it easier to use; for example, in establishing harvesting windows and estimating the dormancy intensity of crops during subsequent winters.

Measuring stress resistance (SR) can be a very laborious process, but an important one because it describes a plant's ability to tolerate the stresses associated with the harvesting-to-outplanting process. Because the seasonal pattern of SR closely coincides with the pattern of cold hardiness, standard cold hardiness tests can provide quick and useful estimates of SR. Moreover, SR is related to dormancy intensity expressed as a dormancy release index. Because refrigerated storage slows the release of bud dormancy, storage prolongs the period of high SR.

Root growth potential (RGP) is the most popular performance test. This test provides an indication of the overall viability of the plant at the time of testing because many integrated physiological processes in plants are responsible for new root production. This test provides only a snapshot-in-time evaluation of the plant; it is important to remember that physiological quality can change right up until the stock is outplanted. RGP may or may not be well correlated with survival because site conditions can override stock quality, but plants having low RGP should be further evaluated with respect to potential site conditions.

In general, morphological attributes, because they seldom change during the harvest-to-outplanting process, may be measured any time. Physiological attributes, because they can change frequently, however, provide only a momentary analysis of plant quality. Testing plant moisture stress at different stages of the harvest-to-outplanting process can ensure that plant stress is minimized. Chlorophyll fluorescence and root electrolyte leakage tests may be used immediately after an unexpected stress event to ascertain damage levels or recovery from those events. Cold hardiness testing can be done to determine proper harvesting windows and prior to outplanting to ensure that stress resistance is still high. Performance attributes such as stress resistance may be done anytime during the harvesting-to-outplanting process, but root growth potential is probably best done immediately prior to outplanting to ensure overall plant viability.

None of these plant quality tests will yield meaningful results unless the population of plants is sampled randomly and thoroughly. Plant producers and users must be cognizant of what each test does and does not infer about plant quality and must be mindful that test results must be considered within the context of expected, but never fully predicted, site conditions.

7.2.10 Literature Cited

Adams, G.T.; Perkins, T.D.; Klein, R.M. 1991. Anatomical studies on first-year winter injured red spruce foliage. American Journal of Botany 78: 1199-1206.

Anderson, J.L.; Seeley, S.D. 1993. Bloom delay in deciduous fruits. In: Janick J., ed. Horticultural Reviews 15: 97-144.

Arnott, J.T.; Beddows, D. 1982. Influence of Stryroblock™ container size on field performance of Douglas-fir, western hemlock and Sitka spruce. Tree Planters' Notes 33(3): 31-34.

Balk, P.A.; Bronnum, P.; Perks, M.; Stattin, E.; van der Geest, L.H.M.; van Wordragen, M.F. 2007. Innovative cold tolerance test for conifer seedlings. In: Riley, L.E.; Dumroese, R.K.; Landis, T.D., tech. coords. National Proceedings: Forest and Conservation Nursery Associations—2006. Proceedings RMRS-P-50. Fort Collins, CO: USDA Forest Service, Rocky Mountain Research Station: 9-12.

Balk, P.A.; Haase, D.L.; van Wordragen, M.F. 2008. Gene activity test determines cold tolerance in Douglas-fir seedlings. In: Dumroese, R.K.; Riley, L.E., tech. coords. National Proceedings: Forest and Conservation Nursery Associations—2007. Proceedings RMRS-P-57. Fort Collins, CO: USDA Forest Service, Rocky Mountain Research Station: 140-148.

Becwar, M.R.; Rajashekar, C.; Bristow, K.J.H.; Burke, M.J. 1981. Deep undercooling of tissue water and winter hardiness limitations in timberline flora. Plant Physiology 68: 111-114.

Bigras, F.J. 2005. Photosynthetic response of white spruce families to drought stress. New Forests 29: 135-148.

Bigras, F.J.; Ryyppo, A.; Lindstrom, A.; Stattin, E. 2001. Cold acclimation and deacclimation of shoots and roots of conifer seedlings. In: Bigras, F.J.; Colombo, S.J., eds. Conifer cold hardiness. Dordrecht, The Netherlands: Kluwer Academic Publishers: 57-88.

Binder, W.D.; Fielder, P.; Mohammed, G.H.; L'Hirondelle, S.J. 1997. Applications of chlorophyll fluorescence for stock quality assessment with different types of fluorometers. New Forests 13: 63-89.

Binder, W.D.; Skagel, R.K.; Krumlik, G.K. 1988. Root growth potential: Facts, myths, value? In: Landis, T.D., ed. Proceedings, combined meeting of the Western Forest Nursery Associations. Gen. Tech. Rep. RM-167. Fort Collins, CO: USDA Forest Service, Rocky Mountain Forest and Range Experiment Station: 111-118.

Burdett, A.N. 1979. New methods for measuring root growth capacity: their value in assessing lodgepole pine stock quality. Canadian Journal of Forest Research 9: 63-67.

Burdett, A.N.; Simpson, D.G. 1984. Lifting, grading, packaging and storing. In: Duryea, M.L.; Landis, T.D., eds. Forest nursery manual: production of bareroot seedlings. The Hague, The Netherlands: Martinus Nijhoff Publishers: 227-234.

Burr, K.E. 1990. The target seedling concepts: bud dormancy and cold hardiness. In: Rose, R.; Campbell, S.J.; Landis, T.D., eds. Target seedling symposium: proceedings, combined meeting of the Western Forest Nursery Associations. Gen. Tech. Rep. RM-200. Fort Collins, CO: USDA Forest Service, Rocky Mountain Forest and Range Experiment Station: 79-90.

Burr, K.E.; Hawkins, C.D.B.; L'Hirondelle, S.J.; Binder, W.D.; George, M.F.; Tapani, R. 2001. Methods for measuring cold hardiness of conifers. In: Bigras, F.J.; Colombo, S.J., eds. Conifer cold hardiness. Dordrecht, The Netherlands: Kluwer Academic Publishers: 369-401.

Burr, K.E.; Tinus, R.W.; Wallner, S.J.; King, R.M. 1989. Relationships among cold hardiness, root growth potential and bud dormancy in three conifers. Tree Physiology 5: 291-306.

Burr, K.E.; Tinus, R.W.; Wallner, S.J.; King, R.M. 1990. Comparison of three cold hardiness tests for conifer seedlings. Tree Physiology 6: 351-369.

Campbell, R.K. 1978. Regulation of bud burst timing by temperature and photoperiod regime during dormancy. In: Hollis, C.A.; Squillace, A.E, eds. Proceedings of fifth North American Forest Biology Workshop. Gainesville, FL: University of Florida, School of Forest Resources and Conservation: 19-34.

Cannell, M.G.R.; Sheppard, L.J. 1982. Seasonal changes in the frost hardiness of provenances of *Picea sitchensis* in Scotland. Forestry 55: 137-153.

Cannell, M.G.R.; Tabbush, P.M.; Deans, J.D.; Hollingsworth, M.K.; Sheppard, L.J.; Phillipson, J.J.; Murray, M.B. 1990. Sitka spruce and Douglas-fir seedlings in the nursery and in cold storage: root growth potential, carbohydrate content, dormancy, frost hardiness and mitotic index. Forestry 63: 9-27.

Chiatante, D.; Di Iorio, A.; Sarnataro, M.; Scippa, G.S. 2002. Improving vigour assessment of pine (*Pinus nigra* Arnold) seedlings before their use in reforestation. Plant Biosystems 136: 209-216.

Colombo, S.J. 2005. The thin green line: a symposium on the state-of-the-art in reforestation. Forest Research Information Paper 160. Sault Saint Marie, ON, Canada: Ontario Ministry of Natural Resources. 175 p.

Colombo, S.J. 2009. Personal communication. Sault Saint Marie, ON, Canada: Ontario Ministry of Natural Resources.

Colombo, S.J.; Sampson, P.H.; Templeton, C.W.G.; McDonough, T.C.; Menes, P.A.; DeYoe, D.; Grossnickle, S.C. 2001. Assessment of nursery stock quality in Ontario. In: Wagner, R.G.; Colombo, S.J. Regenerating the Canadian forest: principles and practice for Ontario. Sault Saint Marie, ON, Canada: Ontario Ministry of Natural Resources: 307-323.

Colombo, S.J.; Zhao, S.; Blumwald, E. 1995. Frost hardiness gradients in shoots and roots of *Picea mariana* seedlings. Scandinavian Journal of Forest Research 9: 1-5.

Coursolle, C.F.; Bigras, J.; Margolis, H.A. 2000. Assessment of root freezing damage of two-year-old white spruce, black spruce and jack pine seedlings. Scandinavian Journal of Forest Research 15: 343-353.

Demig-Adams, B.; Adams, W.W. 1992. Photoprotection and other responses of plants to high light stress. Annual Review of Plant Physiology and Plant Molecular Biology 43: 599-626.

Dexter, S.T.; Tottingham, W.E.; Graber, L.F. 1932. Investigations of the hardiness of plants by measurement of electrical conductivity. Plant Physiology 7: 63-78.

Dixon, H.H.1914. Transpiration and the ascent of sap in plants. New York: MacMillan. 177 p.

Dominguez-Lerena, S.; Herrero Sierra, H.; Carrasco Manzano, I.; Ocaña Bueno, L.; Peñuelas Rubira, J.L.; Mexal, J.G. 2006. Container characteristics influence *Pinus pinea* seedling development in the nursery and field. Forest Ecology and Management 221(1-3): 63-71.

Duryea, M.L. 1985. Evaluating seedling quality: principles, procedures and predictive abilities of major tests. Corvallis, OR: Oregon State University, Forest Research Laboratory. 143 p.

Duryea. M.L.; Landis,T.D., eds. 1984. Forest nursery manual: production of bareroot seedlings. The Hague/Boston/Lancaster: Martinus Nijhoff/Dr W. Junk Publishers: 386 p.

Folk, R.S.; Grossnickle, S.C.; Axelrod, P.; Trotter, D. 1999. Seed lot, nursery, and bud dormancy effects on root electrolyte leakage of Douglas-fir (*Pseudotsuga menziesii*) seedlings. Canadian Journal of Forest Research 29: 1269-1281.

Frampton, L.; Isik, K.; Goldfarb, B. 2002. Effects of nursery characteristics on field survival and growth of loblolly pine rooted cuttings. Southern Journal of Applied Forestry 26: 207-213.

Genty, B.; Briantais, J.M.; Baker, N.R. 1989. The relationship between the quantum yield of photosynthetic electron transport and quenching of chlorophyll fluorescence. Biochemica et Byophysica Acta 990: 97-92.

George, M.F.; Burke, M.J.; Pellett, H.M.; Johnson, A.G. 1974. Low temperature exotherms and woody plant distribution. HortScience 9: 519-522.

Glerum, C. 1976. Frost hardiness of forest trees. In: Cannell, M.G.R.; Last, F.T., eds. Tree physiology and yield improvement. New York: Academic Press: 403-420.

Government of Québec. 2007. Field guide: grading of containerized conifer stock. [Guide terrain : Inventaire de qualification des plants résineux cultivés en récipient.] Québec, QC, Canada: Ministère des Ressources Naturelles et de la Faune, Direction de la production des semences et des plants. 128 p.

Govindjee, R. 1995. Sixty-three years since Kautsky: chlorophyll$_a$ fluorescence. Australian Journal of Plant Physiology 22: 131-160.

Greer, D.H.; Leinonen, I.; Repo, T. 2001. Modelling cold hardiness development and loss in conifers. In: Bigras, F.J.; Colombo, S.J., eds. Conifer cold hardiness. Dordrecht, The Netherlands: Kluwer Academic Publishers: 437-460.

Grossnickle, S.C. 2000. Ecophysiology of northern spruce species: the performance of planted seedlings. Ottawa, ON, Canada: NRC Research Press and National Research Council of Canada. 409 p.

Grossnickle, S.C. 2005. Seedling size and reforestation success: How big is big enough? In: Colombo, S.J., comp. Proceedings, the thin green line: a symposium on the state of the art in reforestation. Forest Research Information Paper 160. Sault Saint Marie, ON, Canada: Ministry of Natural Resources, Ontario Forest Research Institute: 144-149.

Grossnickle, S.C. 2008. Personal communication. Brentwood Bay, British Columbia, Canada: CellFor, Inc.

Grossnickle, S.C.; Folk, R.S. 1993. Stock quality assessment: forecasting survival or performance on a reforestation site. Tree Planters' Notes 44: 113-121.

Grossnickle, S.C.; Major, J.E.; Arnott, J.T.; Lemay, V.M. 1991. Stock quality assessment through an integrated approach. New Forests 5(2): 77-91.

Haase, D.L. 2008. Understanding forest seedling quality: measurements and interpretation. Tree Planters' Notes 52(2): 24-30.

Haase, D.L.; Rose, R. 1995. Vector analysis and its use for interpreting plant nutrient shifts in response to silvicultural treatments. Forest Science 41(1): 54-66.

Harper, C.P.; O'Reilly, C.O. 2000. Effect of warm storage and date of lifting on the quality of Douglas-fir seedlings. New Forests 20: 1-13.

Harrington, J.T.; Mexal, J.D.; Fisher, J.T. 1994. Volume displacement method provides a quick and accurate way to quantify new root production. Tree Planters' Notes 45: 121-124.

Helenius, P.; Luoranen, J.; Rikala, R. 2005. Physiological and morphological responses of dormant and growing Norway spruce container seedlings to drought after planting. Annals of Forest Science 62: 201-207.

Hermann, R.K. 1967. Seasonal variation in sensitivity of Douglas-fir seedlings to exposure of roots. Forest Science 13: 140-149.

Hines, F.D.; Long, J.N. 1986. First and second-year survival of containerized Engelmann spruce in relation to initial seedling size. Canadian Journal of Forest Research 16: 668-670.

Howell, K.D.; Harrington, T.B. 2004. Nursery practices influence seedling morphology, field performance, and cost efficiency of containerized cherrybark oak. Southern Journal of Applied Forestry 28: 152-162.

J.H. Stone Nursery. 1996. Nursery handbook—folder 6075 quality monitoring. Central Point, OR: USDA Forest Service, J.H. Stone Nursery.

Jenkinson, J.L. 1975. Seasonal patterns of root growth capacity in western yellow pines. In: Proceedings, convention of the Society of American Foresters, Washington, D.C., 75th National Convention: 445-453.

Jenkinson, J.L.; Nelson, J.A.; Huddleston, M.E. 1993. Improving planting stock quality—the Humboldt experience. Gen. Tech. Rep. PSW-143. Berkeley, CA: USDA Forest Service, Pacific Southwest Research Station. 219 p.

Jobidon, R.; Charette, L.; Bernier, P.Y. 1998. Initial size and competing vegetation effects on water stress and growth of *Picea mariana* (Mill.) seedlings planted in three different environments. Forest Ecology and Management 103: 293-305.

Jones, G.E.; Cregg, B.M. 2006. Budbreak and winter injury in exotic firs. HortScience 41(1): 143-148.

Kooistra, C.M. 2003. Seedling storage and handling in western Canada. In: Riley, L.E.; Dumroese, R.K.; Landis,T.D., tech. coords. National Proceedings: Forest and Conservation Nursery Associations—2003. Proceedings RMRS-P-33. Ogden, UT: USDA Forest Service, Rocky Mountain Research Station: 15-21.

Krause, G.H.; Weis, E. 1991. Chlorophyll fluorescence and photosynthesis: the basics. Annual Review of Plant Physiology and Plant Molecular Biology 42: 313-349.

Landis, T.D. 1985. Mineral nutrition as an index of seedling quality. In: Duryea, M.L., ed. Evaluating seedling quality: principles, procedures, and predictive abilities of major tests: proceedings of a workshop. Corvallis, OR: Oregon State University, Forest Research Laboratory: 29-48.

Landis, T.D. 2007. Miniplug transplants: producing large plants quickly. In: Riley, L.E.; Dumroese, R.K.; Landis, T.D., tech. coords. National Proceedings: Forest and Conservation Nursery Associations—2006. Proceedings RMRS-P-50. Ogden, UT: USDA Forest Service, Rocky Mountain Research Station: 46-53.

Landis, T.D.; Skagel, R.G. 1988. Root growth potential as an indicator of outplanting performance: problems and perspectives. In: Landis, T.D., ed. Proceedings, combined meeting of the Western Forest Nursery Associations. Gen. Tech. Rep. RM-167. Fort Collins, CO: USDA Forest Service, Rocky Mountain Forest and Range Experiment Station: 106-110.

Landis, T.D.; Haase, D.L.; Dumroese, R.K. 2005. Plant nutrient testing and analysis in forest and conservation nurseries. In: Dumroese, R.K.; Riley, L.E.; Landis, T.D., tech. coords. National proceedings, Forest and Conservation Nursery Associations—2004. Proceedings RMRS-P-35. Fort Collins, CO: USDA Forest Service, Rocky Mountain Research Station: 76-84.

Landis, T.D.; Tinus, R.W.; McDonald, S.E.; Barnett, J.P. 1989. Seedling nutrition and irrigation, vol. 4, the container tree nursery manual. Agric. Handbk. 674. Washington, DC: USDA Forest Service. 119 p.

Lavender, D.P. 1984. Bud dormancy. In: Duryea, M.L., ed. Evaluating seedling quality: principles, procedures, and predictive abilities of major tests. Corvallis, OR: Oregon State University, Forest Research Laboratory: 7-15.

L'Hirondelle, S.J.; Simpson, D.G.; Binder, W.D. 2007. Chlorophyll fluorescence, root growth potential, and stomatal conductance as estimates of field performance potential in conifer seedlings. New Forests 34: 235-251.

Lindqvist, H. 2000. Plant vitality in deciduous ornamental plants affected by lifting date and cold storage. Alnarp, Sweden: Swedish University of Agricultural Sciences. PhD dissertation.

Lindström, A.; Mattsson, A. 1989. Equipment for freezing roots and its use to test cold resistance of young and mature roots of Norway spruce seedlings. Scandinavian Journal of Forest Research 4: 59-66.

Lopushinsky, W. 1990. Seedling moisture status. In: Rose, R.; Campbell, S.J.; Landis, T.D., eds. Proceedings, target seedling symposium, combined meeting of Western Forest Nursery Associations. Gen. Tech. Rep. RM-200. Fort Collins, CO: USDA Forest Service: Rocky Mountain Forest and Range Experiment Station: 123-138.

Lopushinsky, W.; Max, T.A. 1990. Effect of soil temperature on root and shoot growth and on budburst timing in conifer seedling transplants. New Forests 4(2): 107-124.

Maki, D.S.; Colombo, S.J. 2001. Early detection of the effects of warm storage on conifer seedlings using physiological tests. Forest Ecology and Management 154(1-2): 237-249.

Marshall, J.D. 1983. Carbohydrate status as a measure of seedling quality. In: Duryea, M.L., ed. Evaluating seedling quality: principles, procedures, and predictive abilities of major tests: proceedings of a workshop. Corvallis, OR: Oregon State University, Forest Research Laboratory: 49-58.

McCreary, D.; Duryea, M.L. 1984. OSU vigor tests: principles, procedures and predictive ability. In: Duryea, M.L., ed. Evaluating seedling quality: principles, procedures, and predictive abilities of major tests: proceedings of a workshop. Corvallis, OR: Oregon State University, Forest Research Laboratory: 85-92.

McDonald, S.E.; Running, S. 1979. Monitoring irrigation in western forest tree nurseries. Gen. Tech. Rep. RM-61. Fort Collins, CO: USDA Forest Service, Rocky Mountain Forest and Range Experiment Station. 8 p.

McKay, H.H. 1992. Electrolyte leakage from fine roots of conifer seedlings: a rapid index of plant vitality following cold storage. Canadian Journal of Forest Research 22: 1371-1377.

McKay, H.H. 1998. Root electrolyte leakage and root growth potential as indicators of spruce and larch establishment. Silva Fennica 32: 241-252.

McKay, H.H.; Mason, W.L. 1991. Physiological indicators of tolerance to cold storage in Sitka spruce and Douglas-fir seedlings. Canadian Journal of Forest Research 21: 890-901.

McKay, H.H.; Milner, A.D. 2000. Species and seasonal variability in the sensitivity of seedling conifer roots to drying and rough handling. Forestry 73: 259-270.

McKay, H.H.; Morgan, J.L. 2001. The physiological basis for the establishment of bare-root larch seedlings. Forest Ecology and Management 142: 1-18.

McKay, H.H.; White, M.S. 1997. Fine root electrolyte leakage and moisture content: indices of Sitka spruce and Douglas-fir seedling performance after desiccation. New Forests 13: 139-162.

McMinn, R. 1982. Size of container-grown seedlings should be matched to site conditions. In: Scarratt, J.B.; Glerum, C.; Paxman, C.A., eds. Proceedings, Canadian containerized tree seedling symposium, Toronto, Ontario. COJFRC symposium proceedings O-P-10. Sault Saint Marie, ON, Canada: Canadian Forestry Service, Great Lakes Forestry Center: 307-312.

Mena-Petite, A.; Estavillo, J.M.; Duñabeitia, M.; González-Moro, B.; Muñoz-Rueda, A.; Lacuesta, M. 2004. Effect of storage conditions on post planting water status and performance of *Pinus radiata* D. Don stock-types. Annals of Forest Science 61: 695-704.

Mena-Petite, A.; Ortega-Lasuen, U.; González-Moro, M.B.; Lacuesta, M.; Muñoz-Rueda, A. 2001. Storage duration and temperature effect on the functional integrity of container and bare-root *Pinus radiata* D. Don seedlings. Trees 15: 289-296.

Mena-Petite, A.; Robreto, A.; Alcalde, S.; Duñabeitia, M.K.; González-Moro, M.B.; Lacuesta, M.; Muñoz-Rueda, A. 2003. Gas exchange and chlorophyll fluorescence responses of *Pinus radiata* D. Don seedlings during and after several storage regimes and their effects on post-planting survival. Trees 17: 133-143.

Mexal, J.G.; Landis, T.D. 1990. Target seedling concepts: height and diameter. In: Rose, R.; Campbell, S.J.; Landis, T.D., eds. Proceedings, target seedling symposium, combined meeting of Western Forest Nursery Associations. Gen. Tech. Rep. RM-200. Fort Collins, CO: USDA Forest Service, Forest and Range Experiment Station: 17-35.

Mohammed, G.H.; Binder, W.D.; Gillies, S.L. 1995. Chlorophyll fluorescence: a review of its practical forestry applications and instrumentation. Scandinavian Journal of Forest Research 10: 383-410.

Öquist, G.; Gardeström, P.; Huner, N.P.A. 2001. Metabolic changes during cold acclimation and subsequent freezing and thawing. In: Bigras, F.J.; Colombo, S.J., eds. Conifer cold hardiness. Dordrecht, The Netherlands: Kluwer Academic Publishers: 137-163.

O'Reilly, C.; McCarthy, N.; Keane, M.; Harper, C.P.; Gardiner, J.J. 1999. The physiological status of Douglas-fir seedlings and the field performance of freshly lifted and cold stored stock. Annals of Forest Science 56: 297-306.

Palta, J.P.; Levitt, J.; Stadlemann, E.J. 1977. Freezing injury in onion bulb cells. I. Evaluation of the conductivity method and analysis of ion and sugar efflux from injured cells. Plant Physiology 60: 393-397.

Peguero-Pina, J.J.; Morales, F.; Gil-Pelegrin, E. 2008. Frost damage in *Pinus sylvestris* L. stems assessed by chlorophyll fluorescence in cortical bark chlorenchyma. Annals of Forest Science 65(813). 6 p.

Perks, M.P.; Monaghan, S.; O'Reilly, C.; Osborne, B.A.; Mitchell, D.T. 2001. Chlorophyll fluorescence characteristics, performance and survival of freshly lifted and cold stored Douglas-fir seedlings. Annals of Forest Science 58: 225-235.

Perry, K. 1998. Basics of frost and freeze protection for horticultural crops. HortTechnology 8: 10-15.

Perry, T.O. 1971. Dormancy of trees in winter. Science 171: 29-36.

Puttonen, P. 1986. Carbohydrate reserves in *Pinus sylvestris* seedling needles as an attribute of seedling vigor. Scandinavian Journal of Forest Research 1(2): 181-193.

Quamme, H.A. 1985. Avoidance of freezing injury in woody plants by deep supercooling. Acta Horticultura 168: 11.

Richardson, E.A.; Seeley, S.D.; Walker, D.R. 1974. A model for estimating the completion of rest for "Redhaven" and "Elberta" peach trees. HortScience 9: 331-332.

Rietveld, W.J.; Tinus, R.W. 1990. An integrated technique for evaluating root growth potential of tree seedlings. Research Note RM-497. Fort Collins, CO: USDA Forest Service, Rocky Mountain Forest and Range Experiment Station. 11 p.

Ritchie, G.A. 1984a. Effect of freezer storage on bud dormancy release in Douglas-fir seedlings. Canadian Journal of Forest Research 14: 186-190.

Ritchie, G.A. 1984b. Assessing seedling quality. In: Duryea. M.L.; Landis,T.D., eds. Forest nursery manual: production of bareroot seedlings. The Hague/Boston/Lancaster: Martinus Nijhoff/Dr W. Junk Publishers: 243-259.

Ritchie, G.A. 1985. Root growth potential: principles, procedures and predictive ability. In: Duryea, M.L, ed. Evaluating seedling quality: principles, procedures, and predictive abilities of major tests. Corvallis, OR: Oregon State University, Forest Research Laboratory: 93-104.

Ritchie, G.A. 1986. Relationships among bud dormancy status, cold hardiness, and stress resistance in 2+0 Douglas-fir. New Forests 1: 29-42.

Ritchie, G.A. 1989. Integrated growing schedules for achieving physiological uniformity in coniferous planting stock. Forestry (Suppl) 62: 213-226.

Ritchie, G.A. 1991. Measuring cold hardiness. In: Lassoie, J.P.; Hinckley, T.M., eds. Techniques and approaches in forest tree ecophysiology. Boca Raton, FL: CRC Press: 557-582.

Ritchie, G.A. 2000. The informed buyer: understanding seedling quality. In: Rose, R.; Haase, D.L., eds. Conference proceedings, advances and challenges in forest regeneration, Nursery Technology Cooperative, Oregon State University and Western Forestry and Conservation Association: 51-56.

Ritchie, G.A.; Dunlap, J.R. 1980. Root growth potential: its development and expression in forest tree seedlings. New Zealand Journal of Forest Science 10: 218-248.

Ritchie, G.A.; Hinckley, T.M. 1975. The pressure chamber as an instrument for ecological research. Advances in Ecological Research 9: 165-254.

Ritchie, G.A.; Shula, R.G. 1984. Seasonal changes of tissue-water relations in shoots and root systems of Douglas-fir seedlings. Forest Science 30: 538-548.

Ritchie, G.A.; Tanaka, Y. 1990. Root growth potential and the target seedling. In: Rose, R.; Campbell, S.J.; Landis, T.D., eds. Proceedings, target seedling symposium, combined meeting of Western Forest Nursery Associations. Gen. Tech. Rep. RM-200. Fort Collins, CO: USDA Forest Service, Rocky Mountain Forest and Range Experiment Station: 37-51.

Ritchie, G.A.; Roden, J.R.; Kleyn, N. 1985. Physiological quality of lodgepole pine and interior spruce seedlings: effects of lift date and duration of freezer storage. Canadian Journal of Forest Research 15: 636-645.

Ronco, F. 1973. Food reserves of Engelmann spruce planting stock. Forest Science 19: 213-219.

Rose, R.; Haase, D.L. 2002. Chlorophyll fluorescence and variations in tissue cold hardiness in response to freezing stress in Douglas-fir seedlings. New Forests 23: 81-96.

Rose, R.; Haase, D.L.; Kroiher, F.; Sabin, T. 1997. Root volume and growth of ponderosa pine and Douglas-fir seedlings: a summary of eight growing seasons. Western Journal of Applied Forestry 12: 69-73.

Sakai, A.; Weiser, C.J. 1973. Freezing resistance of trees in North America with reference to tree regions. Ecology 54: 118-126.

Scholander, P.F.; Hammel, H.T.; Bradstreet, E.D.; Hemmingson, E.A. 1965. Sap pressure in vascular plants. Science 148: 339-346.

Schreiber, U.; Bilger, W.; Neubauer, C. 1995. Chlorophyll fluorescence as a nonintrusive indicator of rapid assessment of in vivo photosynthesis. In: Schultze, E.O.; Caldwell, M.M., eds. Ecophysiology of Photosynthesis. Berlin, Heidelberg, New York: Springer-Verlag: 48-70.

Simpson, D.G. 1990. Frost hardiness, root growth capacity, and field performance relationships in interior spruce, lodgepole pine, Douglas-fir, and western hemlock seedlings. Canadian Journal of Forest Research 20: 566-572.

Simpson, D.G.; Ritchie, G.A. 1997. Does RGP predict field performance? A debate. New Forests 13: 253-277.

Slatyer, R.O. 1967. Plant water relationships. London and New York: Academic Press: 366 p.

Sorensen, F.C. 1983. Relationship between logarithms of chilling period and germination or bud flush rate is linear for many tree species. Forest Science 29: 237-240.

South, D.B.; Mitchell, R.G. 2006. A root-bound index for evaluating planting stock quality of container-grown pines. Southern African Forestry Journal 207: 47-54.

Stattin, E.; Hellqvist, C.; Lindström, A. 2000. Storability and root freezing tolerance of Norway spruce (*Picea abies*) seedlings. Canadian Journal of Forest Research 30: 964-970.

Stone, E.C. 1955. Poor survival and the physiological condition of planting stock. Forest Science 1: 90-94.

Sundheim, I.; Kohmann, K. 2001. Effects of thawing procedure on frost hardiness, carbohydrate content and timing of bud break in *Picea abies*. Scandinavian Journal of Forest Research 16: 30-36.

Sutinen, M.L.; Arora, R.; Wisniewski, M.; Ashworth, E.; Strimbeck, R.; Palta, J. 2001. Mechanisms of frost survival and freeze-damage in nature. In: Bigras, F.J.; Colombo, S.J., eds. Conifer cold hardiness. Dordrecht, The Netherlands: Kluwer Academic Publishers: 89-120.

Sutton, R.F. 1983. Root growth capacity: relationship with field root growth and performance in outplanted jack pine and black spruce. Plant and Soil 71: 111-122.

Tanaka, Y.; Brotherton, P.; Hostetter, S.; Chapman, D.; Dyce, S.; Belanger, J.; Johnson, B.; Duke, S. 1997. The operational planting stock quality testing program at Weyerhaeuser. New Forests 13: 423-437.

Thiffault, N. 2004. Stock type in intensive silviculture: a (short) discussion about roots and size. Forestry Chronicle 80(4): 463-468.

Thompson, B.E. 1985. Seedling morphological evaluation: what you can tell by looking. In: Duryea, M.L., ed. Evaluating seedling quality: principles, procedures, and predictive abilities of major tests. Corvallis, OR: Oregon State University, Forest Research Laboratory: 59-71.

Timmer, V.R. 1997. Exponential nutrient loading: a new fertilization technique to improve seedling performance on competitive sites. New Forests 13: 279-299.

Timmis, K.A.; Fuchigami, L.H.; Timmis, R. 1981. Measuring dormancy: the rise and fall of square waves. HortScience 16: 200-202.

Timmis, R. 1976. Methods of screening tree seedlings for frost hardiness. In: Cannell, M.G.R.; Last, F.T., eds. Tree physiology and yield improvement. London and New York: Academic Press: 421-435.

Timmis, R.; Tanaka, Y. 1976. Effects of container density and plant water stress on growth and cold hardiness of Douglas-fir seedlings. Forest Science 22(2): 167-172.

Timmis, R.; Worrall, J. 1975. Environmental control of cold acclimation in Douglas-fir during germination, active growth and rest. Canadian Journal of Forest Research 5: 464-477.

Toivonen, A.; Rikala, R.; Repo, P.; Smolander, H. 1991. Autumn colouration of first year *Pinus sylvestris* seedlings during frost hardening. Scandinavian Journal of Forest Research 6(1): 31-39.

van den Driessche, R. 1977. Survival of coastal and interior Douglas-fir seedlings after storage at different temperatures, and effectiveness of cold storage in satisfying chilling requirements. Canadian Journal of Forest Research 7: 125-131.

van den Driessche, R. 1984. Relationship between spacing and nitrogen fertilization of seedlings in the nursery, seedling mineral nutrition, and outplanting performance. Canadian Journal of Forest Research 14: 431-436.

van den Driessche, R. 1987. Importance of current photosynthate to new root growth in planted conifer seedlings. Canadian Journal of Forest Research 17: 776-782.

van den Driessche, R. 1991. New root growth of Douglas-fir seedlings at low carbon dioxide concentration. Tree Physiology 8: 289-295.

Vidaver, W.; Toivonen, P.; Lister, G.; Brooke, R.; Binder, W. 1988. Variable chlorophyll-A fluorescence and its potential use in tree seedling production and forest regeneration. In: Landis, T.D., ed. Proceedings, combined meeting of the Western Forest Nursery Associations. Gen. Tech. Rep. RM-167. Fort Collins, CO: USDA Forest Service, Rocky Mountain Forest and Range Experiment Station: 127-132.

Vidaver, W.E.; Lister, G.R.; Brooke, R.C.; Binder, W.D. 1991. A manual for the use of variable chlorophyll fluorescence in the assessment of the ecophysiology of conifer seedlings. FRDA Report 163, British Columbia, Canada. 65 p.

Wakeley, P.C. 1949. Physiological grades of southern pine nursery stock. In: Shirley, H.L., ed. Proceedings, Society of American Foresters Annual Meeting: 311-321.

Wakeley, P.C. 1954. Planting the southern pines. Agricultural Monograph No. 18. Washington, DC: USDA Forest Service. 233 p.

Waring, R.H.; Cleary, B.D. 1967. Plant moisture stress: evaluation by pressure bomb. Science 155: 1248-1254.

Weiser, C.J. 1970. Cold resistance and injury in woody plants. Science 169: 1269-1278.

Wenny, D.L.; Swanson, D.J.; Dumroese, R.K. 2002. The chilling optimum of Idaho and Arizona ponderosa pine buds. Western Journal of Applied Forestry 17: 117-121.

Wheeler, N.C.; Jermstad, K.D.; Krutovsky, K.; Aitken, S.N.; Howe, G.T.; Krakowski, J.; Neale, D.B. 2005. Mapping of quantitative trait loci controlling adaptive traits in coastal Douglas-fir. IV. Cold-hardiness QTL verification and candidate gene mapping. Molecular Breeding 15: 145-156.

Wilner, J. 1955. Results of laboratory tests for winter hardiness of woody plants by electrolyte methods. Proceedings of the American Horticultural Society 66: 93-99.

Wilner, J. 1960. Relative and absolute electrolyte conductance tests for frost hardiness of apple varieties. Canadian Journal of Plant Science 40: 630-637.

Wilson, B.C.; Jacobs, D.F. 2006. Quality assessment of temperate and deciduous hardwood seedlings. New Forests 31: 417-433.

7.2.11 Appendix

Appendix 7.2.1—*Seedling quality testing facilities and their procedures*

Company	Address	Types of tests offered			
		Morphology	Root growth potential	Cold hardiness	Others
Nursery Technology Cooperative	Oregon State University Dept. of Forest Science Richardson Hall 321 Corvallis, OR 97331 Tel: 541-737-6576 Fax: 541-737-1393 http://ntc.forestry.oregonstate.edu/sqes	X		X	
KBM Forestry Consultants	SQA Coordinator 349 Mooney Avenue Thunder Bay, ON P7B 5L5 Tel: 807-345-5445 ext. 34 Fax: 807-345-3440 E-mail: sgellert@kbm.on.ca	X	X	X	X
Laboratory for Forest Soils and Environmental Quality	Tweeddale Centre for Industrial Forest Research 1350 Regent Street Fredericton, NB E3C 2G6 Tel: 506-458-7817 Fax: 506-453-3574 E-mail: jestey@unb.ca	X	X	X	X
Franklin H. Pitkin Nursery	Center for Forest Nursery and Seedling Research College of Natural Resources University of Idaho Moscow, ID 83844–1137 Tel: 208-885-7023 Fax: 208-885-6226 E-mail: seedlings@uidaho.edu	X	X	X	X

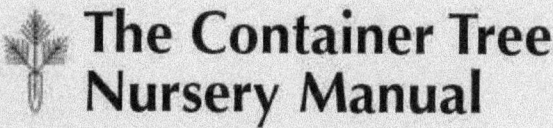

The Container Tree Nursery Manual

Volume Seven

Chapter 3
Harvesting

Contents

7.3.1 Introduction *85*
7.3.1.1 Hot-planting *85*
7.3.1.2 Dormant stock *86*

7.3.2. Scheduling the Winter Harvesting Window *87*
7.3.2.1 Calendar and visual clues *87*
       Foliage characteristics
       Buds
       Presence of white root tips
7.3.2.2 Outplanting trials *88*
7.3.2.3 Plant quality tests *89*
       Estimating bud dormancy with chilling sums
       Cold hardiness testing

7.3.3 Prestorage Fungicide Treatments *90*

7.3.4 Processing Speculation and Contract Crops *91*
7.3.4.1 Small speculation orders *91*
7.3.4.2 Large contract orders *91*

7.3.5 Grading and Packaging *92*
7.3.5.1 Storage and shipping in growth container *92*
7.3.5.2 Plant extraction *92*
7.3.5.3 Packaging plants *94*
       Jellyrolling
       Bagging and boxing
       Boxing
7.3.5.4 Processing large-volume container stock *96*

7.3.6 Packaging for Storage and Shipping *99*

7.3.7 Processing Cull Seedlings *100*

7.3.8 Summary and Conclusions *101*

7.3.9 Literature Cited *102*

7.3.1 Introduction

Container nursery managers wait anxiously until they can begin to harvest their crop because, after plants are graded and placed into storage, the value of the crop is at its maximum. Scheduling the best time to harvest is critical because plants need to be at their peak of quality yet hardy enough to withstand the sequential stresses of packing, storage, shipping, and outplanting.

"Lifting" is a historical term adopted from bareroot nursery harvesting when seedlings are physically removed from the soil; the term is still used in container nurseries as an operational synonym for harvesting. "Lift and pack" is another bareroot nursery term that has been adopted by container growers when referring to the harvest operation.

When scheduling plant harvest, the first and most important consideration is whether the stock will be lifted and outplanted immediately ("hot-planted") or harvested when dormant and then stored for later shipment and outplanting.

Methods of harvesting container stock across North America are a function of nursery size and location, plant species, research input, and tradition. Many large nurseries in the Western United States and Canada remove plants from containers and package them ("lift and pack" or "pull and wrap"). They use refrigerated storage to handle large orders that must all be processed simultaneously. This is the case in much of the Pacific Northwest, where winter temperatures are variable and snow is absent or intermittent (for example, Kooistra 2004). In eastern Canada, however, temperatures remain cold enough that container stock can be stored outdoors or some nurseries use snowmaking equipment to supplement the snowpack (White 2004). The harvest and storage of other native plants can be considerably different from commercial conifers. Because of the sheer number of species, the wide variety of container sizes, and the fact that little or no research has been done on dormancy or hardiness, native plants may require special harvest and storage procedures (Burr 2005).

7.3.1.1 Hot-planting

Hot-planting is done during summer or fall, when plants are not completely dormant or hardy; plants must be handled with care throughout the process. This means that stock is lifted, held for a short time with or without refrigerated storage, and outplanted within a week or two. Greenhouse stock that will be hot-lifted is usually held for several weeks in a shadehouse or open compound to develop some degree of hardiness before outplanting (fig. 7.3.1). Some nurseries use moisture stress and/or artificially reduce day length ("blackout") to hasten the hardening process. (More information on hardening nursery stock can be found in Section 6.4.4 of Volume Six and on blackout in Section 3.3.4.6 in Volume Three of this series.)

Timing is the key to a successful hot-planting program, and it is critical to minimize the time from when stock is harvested to when it is outplanted. This tight timeline and the fact that the stock is not fully hardy mean that most hot-plantings must be relatively close to the nursery.

When the customer notifies the nursery that the outplanting sites are ready, the stock is graded to specifications and a final "shippable" inventory is calculated. Plants should be packed standing up in boxes to encourage air exchange and allow for possible irrigation on the outplanting site. Do not pack in plastic liners that restrict air flow and can trap the heat generated by plant respiration. Packed stock

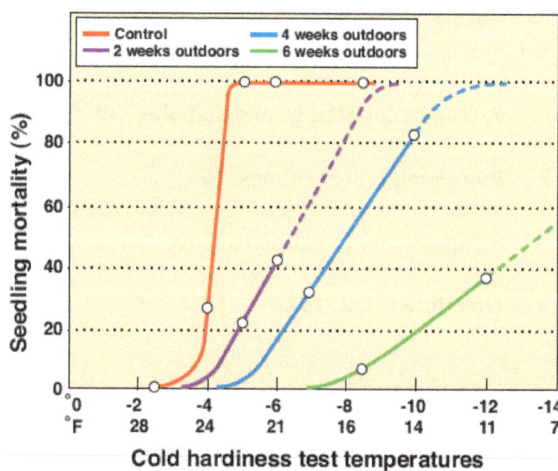

Figure 7.3.1—*All nursery stock must be properly hardened before outplanting, especially those being "hot-planted." The longer these loblolly pine seedlings were hardened in an outdoor compound the more cold hardy they became (modified from Mexal and others 1979).*

should be placed immediately in cooler storage at approximately 4.4 °C (40 °F) (Fredrickson 2003).

For larger planting projects, stock is held for a short period in refrigerated shipping vans at the nursery until the entire order can be shipped. With southern pines, seedlings harvested for summer hot-planting are generally stored at 4 to 21 °C (40 to 70 °F) for a week or less (Dumroese and Barnett 2004). (More information on hot-planting in summer and fall outplanting windows can be found in Section 7.1.2.5.)

7.3.1.2 Dormant stock

Most container nursery stock is harvested during the winter dormant season and stored until it can be shipped for outplanting. Storage methods are discussed in Chapter 7.4. The key consideration in harvesting is whether plants will be placed in open, sheltered, or refrigerated storage. The type of storage dictates not only the time of harvest but also the type of packaging. In open compounds and sheltered storage, plants remain in their containers, whereas, for refrigerated storage, they are typically removed from their containers, graded, and packed in cardboard boxes.

7.3.2 Scheduling the Winter Harvesting Window

Nursery managers must harvest their crop at its peak of quality and know how to maintain that quality until plants are delivered to the customer. This means harvesting when plants are fully dormant and resistant to the stresses of harvesting, storage, shipping, and outplanting. This time period is known as the "lifting window" or "harvesting window."

Foresters and other nursery customers have observed that stock harvested during winter dormancy survive and grow better than plants lifted a few months earlier or later. Numerous inhouse studies and research trials have confirmed these observations. Although most of this research was done with bareroot stock, the same principles apply to container stock. Whereas harvesting bareroot plants at their peak dormancy is usually restricted or compromised by too muddy or frozen soils, container plants can be harvested throughout the winter dormant season. Given this potentially wide harvest season for container stock, let's discuss some ways that growers determine the proper lifting window.

7.3.2.1 Calendar and visual clues

Using calendars and visual clues are the most traditional techniques for scheduling and harvesting, and when based on the combined experience of the nursery staff, can be quite effective. The procedure is simple—if it takes 4 weeks to harvest the crop, then that amount of time is subtracted from when the crop becomes fully dormant or is scheduled to be shipped for outplanting. One calendar technique for deciding when to harvest is known as the "F-date," which is based on the average date of the first fall frost. Harvesting can begin 30 to 45 days after this date (Mathers 2000).

Experienced growers also use several morphological indicators to help them confirm when plants are becoming dormant and hardy and ready to harvest.

Foliage characteristics. Determining when deciduous species are ready to lift is relatively easy, because their leaves change color and eventually fall off. Even evergreen species can show foliar signs when they are becoming dormant. For example, the cuticle of leaves or needles becomes thicker and waxier so that the plant can tolerate desiccation during the winter. Experienced growers can feel a difference in foliage texture and rigidity when plants become hardy, and the needles of some

Figure 7.3.2—*Plants develop visible signs of dormancy and hardiness, such as bluish wax deposits on their foliage (A). Bud size and development are also signs of dormancy and plant quality; large buds containing many leaf primordia (B) are superior to smaller, less developed buds (C) (B&C, courtesy of Steve Colombo).*

species exhibit a slight change in color. For example, the actively growing foliage of Engelmann spruce (*Picea engelmanii*) is bright green, whereas dormant foliage becomes bluer in color because of the waxy cuticle that develops on the surface (fig. 7.3.2A).

Buds (presence, size, and number of primordia). Plants with determinant growth patterns, such as pines and spruces, form a bud at the end of the growing season. In the temperate zone, most people look for large buds with firm scales as an indication of shoot dormancy and plant quality. Other plants, such as junipers and cedars, have indeterminant growth and a terminal bud is not formed. Some semitropical pines, such as longleaf (*Pinus palustris*) in the southern United States, also do not form buds in the nursery (Jackson and others 2007). (See section 6.1 in Volume Six for more information on determinant and indeterminant growth patterns).

Bud size and length have traditionally been used as good indicators of when plants are ready to harvest. In Eastern Canada, counting the number of bud primordia is one way that nurseries determine the timing of their harvests (fig. 7.3.2 B&C). KBM Forestry Consultants, a private seedling testing laboratory in Thunder Bay, Ontario, offers bud dissection on a fee basis (Colombo and others 2001).

Presence of white root tips. Some growers consider the presence or absence or white root tips as a sign of plant dormancy. Roots never go truly dormant, however, and will grow whenever temperatures are favorable. Therefore, the presence of white root tips has little value in predicting dormancy or hardiness, but numerous long, white roots indicate that plants have been exposed to temperatures above 10 °C (50 °F) (see figure 7.2.41C).

So, although it is not too scientific, scheduling the lifting window by the calendar and visual clues can be effective if based on actual nursery and field experience with specific plant species.

7.3.2.2 Outplanting trials

Another traditional method of determining when it is safe to lift nursery stock involves outplanting performance. Over a period of years, nurseries can determine their lifting windows from observations on plant survival and growth after outplanting. This technique has been used for bareroot stock, but few results have been published for container plants. In a comprehensive study of four bareroot conifer seedlings from northern California, samples were collected at monthly intervals throughout the winter and then outplanted and evaluated for first-year survival (Jenkinson and others 1993). The resulting data show that lifting windows can vary significantly between species and between seedlots within a species (fig. 7.3.3). Outplanting trials are effective in establishing lifting windows, but the drawback is that it takes from 5 to 10 years to accumulate enough data to account for seasonal weather variation. In addition, separate trials would be needed for customers from different climatic regions.

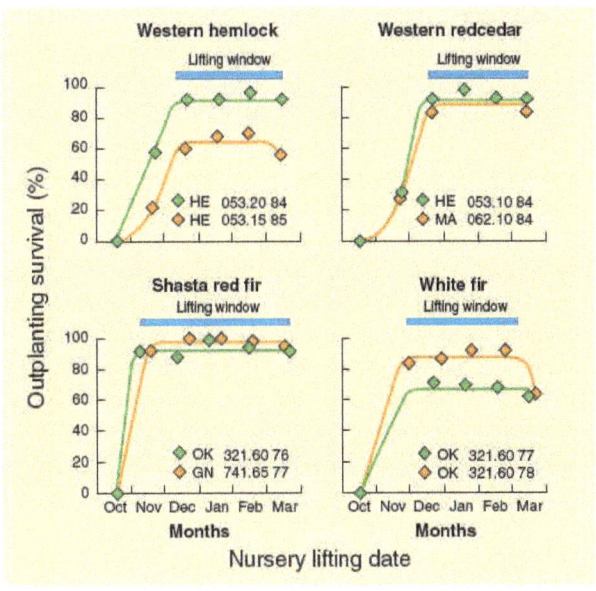

Figure 7.3.3—*An effective but time-consuming way to establish lifting windows is to lift plants throughout the harvest season and monitor their performance after outplanting. These results show that the lifting window for all four conifers was from late November to the end of February. Because they became dormant sooner, high elevation species such as Shasta red fir (*Abies magnifica*) had a wider window than species from lower elevations (modified from Jenkinson and others 1993).*

7.3.2.3 Plant quality tests

Plant quality testing is discussed in detail in Chapter 7.2, and several tests have been used to determine when container stock is ready to harvest. Root growth potential (RGP) is the most widely known quality test, and many experiments have attempted to correlate RGP with lifting windows. Although this test provides an indication of vitality and relative vigor, RGP readings typically vary too much from year to year to be useful.

Estimating bud dormancy with chilling sums. All growers know that plants should be harvested when they are dormant. Unfortunately, growers lack a quick-and-easy test to determine dormancy status—current tests measure only bud dormancy. Therefore, the easiest and most practical method for estimating intensity of bud dormancy is based on the chilling requirement. The concept is logical enough—the cumulative exposure of plants to cold temperatures controls the release of dormancy. Therefore, by measuring the duration of this exposure, it is possible to estimate the intensity of dormancy indirectly.

The operational application is known as chilling sums, or degree-hardening days. The process involves measuring the temperature throughout the day and calculating the duration of time below some reference temperature. This chilling sum can be calculated with several different formulas, and environmental monitoring equipment is available that will calculate chilling sums automatically (see Section 7.2.5.1 for more details).

Cold hardiness testing. It is traditional knowledge that plants should be hardy enough to withstand the stresses of harvesting, storage, shipping, and outplanting. There are many types of hardiness, but cold hardiness has proved to be easiest to measure and the best predictor of when to lift container stock. For more than 20 years, cold hardiness tests have been used to determine lifting windows and estimate storability in Canadian container nurseries. Their critical threshold is when plants exposed to cold temperatures show less than 25 percent visible cold injury to the foliage (Burdett and Simpson 1984). For open storage, plants must be able to tolerate two consecutive frost hardiness tests at –15 °C (5 °F) (Colombo and others 2001; White 2004), whereas for long-term freezer storage, tolerating two consecutive hardiness tests at –15 °C (5 °F) or one at –40 °C (–40 °F) is considered adequate (Colombo and Gellert 2002).

Growth chamber measurements of cold hardiness of Douglas-fir and ponderosa pine container seedlings were modeled against weather data to establish lifting windows (Tinus 1996). Freeze-induced electrolyte leakage tests demonstrated the year-to-year variation in lifting windows that can be expected. Comparing the four years modeled, the starting and ending dates and the duration of the lifting window were significantly different (fig. 7.3.4).

7.3.3 Prestorage Fungicide Treatments

Storage molds are a serious concern during overwinter storage, especially from the fungus *Botrytis cinerea* that is commonly found on lower senescent foliage (fig. 7.3.5A). Therefore, just prior to packing plants for refrigerated storage, some nurseries treat their stock with foliar fungicides.

Unfortunately, both nursery workers and tree planters have complained about skin rashes and other allergic symptoms after handling fungicide-treated stock. The only comprehensive study of fungicide effectiveness and pesticide residue was conducted at container nurseries in British Columbia (Trotter and others 1992). Two fungicides, benomyl (Benlate 50WP) and captan (Captan 50WP), were sprayed on conifer seedlings prior to refrigerated storage using an irrigation boom. In one treatment, both pesticides were applied through a backpack sprayer. They found both fungicides were effective when applied to species predisposed to Botrytis mold. Seedling samples were collected before and after the standard storage period to determine pesticide residue levels. Captan was found to be more persistent than benomyl, and levels were significantly higher when a backpack sprayer was used (fig. 7.3.5B). This short residual effect means that the fungicide is effective only immediately after application and that highly susceptible seedlots may still be infected if predisposing conditions exist during or after storage (Trotter and others 1992).

Therefore, the decision on whether to apply protective fungicides to control storage molds should be considered from both cultural and safety standpoints. Species and seedlots that are already infected prior to harvest may benefit from protective fungicides, but heavily infested or stressed lots may still develop mold problems during or after storage. (See Volume 5, Section 5.1.6.2 for more discussion of molds and other storage problems.)

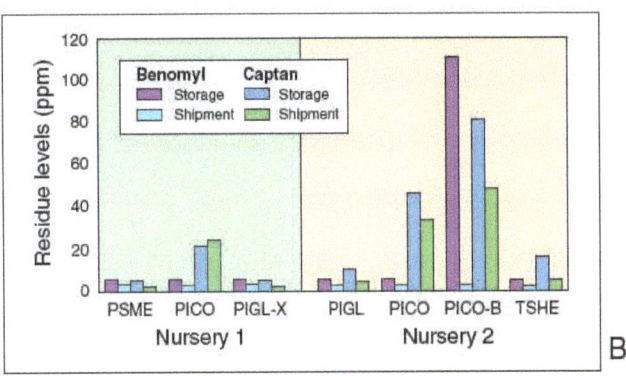

Figure 7.3.5—*Plants grown at high densities often develop Botrytis mold on the lower senescent foliage (A); pesticide residue levels on foliage of stored plants varied by nursery and by application method. All treatments were applied in the greenhouse through an irrigation boom except PICO-B in nursery 2, which was applied with a backpack sprayer (B, species codes: PSME = Pseudotsuga menziesii, PICO = Pinus contorta, PIGL = Picea glauca, PIGL-X = Picea glauca x engelmannii, TSHE = Tsuga heterophylla) (modified from Trotter and others 1992).*

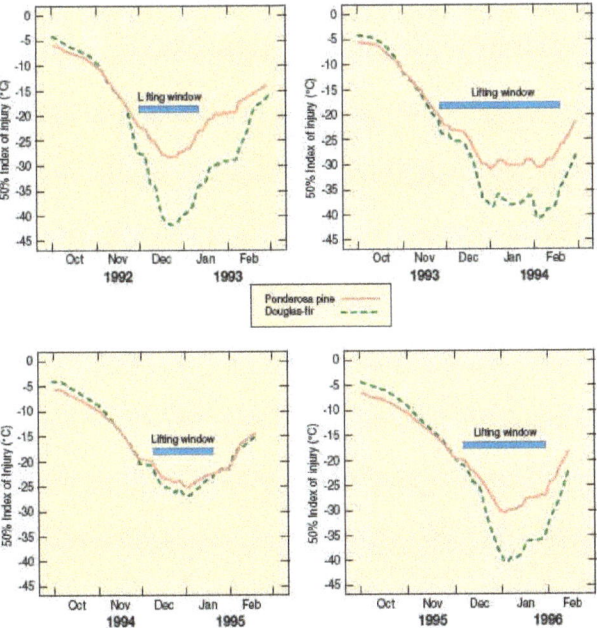

Figure 7.3.4—*Cold hardiness, measured as the 50-percent index of injury from freeze-induced electrolyte leakage tests, was used to model the lifting window for two southwestern conifers over four winters from 1992 to 1996 (modified from Tinus 1996).*

7.3.4 Processing Speculation and Contract Crops

The manner in which container stock is processed depends on how the crops will be sold and shipped.

7.3.4.1 Small speculation orders

Some nurseries, such as State government and private nurseries, service many, perhaps thousands, of customers who order a few plants of many different species. Plants to satisfy these orders are usually grown on speculation, and orders are accepted throughout the winter and spring shipping season. To facilitate filling and processing orders, plants are usually harvested, graded, packaged into discrete quantities (for example, 5 or 25; usually the minimum number that can be ordered), and then stored in bulk bins in a cooler. As orders come in, workers pull plants from the bulk bins and combine the various species for shipping, often by mail or parcel service.

7.3.4.2 Large contract orders

Many Federal government and forest industry nurseries grow all or most of their container plants on contract and so grade, package (often 100 to 500 plants per box), and store their stock in the same operation by customer. Depending on the customer preference and the length of storage, these orders may be held in cooler or freezer storage.

7.3.5 Grading and Packaging

Regardless of whether container plants are going to be hot-planted or outplanted as dormant stock, they are graded for size and appearance according to established standards or, in the case of contract stock, standards agreed to with the customer (see Volume 1, pages 147–149). "Culls" are plants that do not meet the grading criteria. Sometimes these criteria are adjusted during the grading process based on other cull and shipping factors that become apparent during the process. Typical grading criteria include shoot height, stem diameter at the root collar ("caliper"), and root plug integrity (fig. 7.3.6). In addition, plants are inspected for physical injury or disease, especially for gray mold (*Botrytis cinerea*) that can spread in storage.

The timing of the grading operation depends on the harvesting methods. To minimize volume and reduce disease during storage, most container nurseries grade their stock as part of the harvesting process. Some nurseries that store in open compounds ship ungraded stock to the outplanting site, where they are graded immediately before outplanting (Dionne 2006).

Container size and the way plants will be packaged and stored determine the best processing system. For smaller volume containers, plants can be processed two ways: (1) grading, storing, and shipping plants in the growth container; or (2) extracting ("pulling" or "lifting") plants from the growth container and subsequently grading, packaging, and placing them into storage and/or shipping containers (Landis and McDonald 1981). Because of their size and weight, larger volume single containers are graded and handled individually.

7.3.5.1 Storage and shipping in growth container

This process is generally limited to container types with individual soft plastic "cells" or "tubes" that are held in hard plastic racks. The most popular containers of this type are Ray Leach "Cone-tainers"™ and Deepots™ (fig. 7.3.7A). The harvesting process consists of removing each container from the rack, grading the plant within it, and then placing the container into either a "shippable" or "cull" rack (fig. 7.3.7B). Racks of shippable plants are stored outside either in shadehouses or under white plastic sheeting (see Section 7.4) until they can be shipped for outplanting. Cull racks are emptied as time permits. Because the plastic racks are brittle and can be

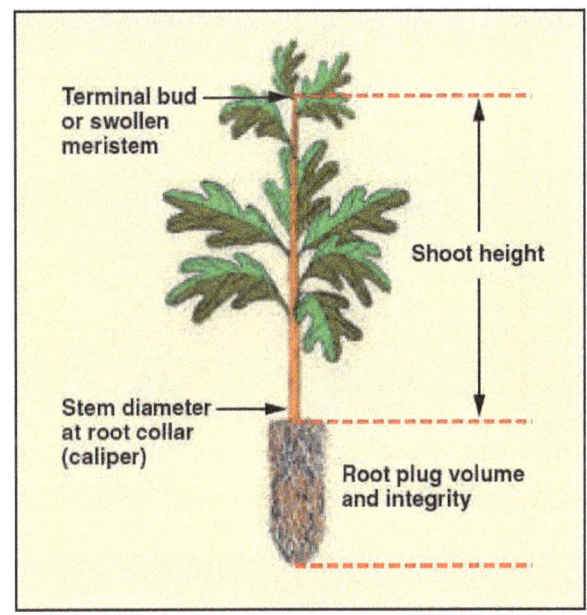

Figure 7.3.6—*Common grading standards include shoot height, stem diameter at the root collar ("caliper"), and root plug integrity.*

damaged during handling and shipping (fig. 7.3.7C), some nurseries group shippable containers into bunches secured by a rubberband or put them in plastic bags (fig. 7.3.7D), which are then placed in cardboard boxes for refrigerated storage.

Some nurseries using block containers also leave graded plants in the growth containers for storage and shipping to the outplanting site. This is particularly common with the more durable hard plastic blocks, such as the Hiko™ tray, IPL Rigi-Pots™, and Ropak Multi-Pots™. In some nurseries, cull plants are pulled from containers during the grading process but, in others, no grading occurs and all plants are shipped to the outplanting site, where the planter makes the final decision regarding plant quality (Dionne 2006).

7.3.5.2 Plant extraction

As mentioned earlier, container plants are typically extracted from their growth container when they will be stored under refrigeration, but often hot-planted stock is

Figure 7.3.7—Harvesting nursery stock in the growth container is popular for soft plastic cells that can be removed from their racks (A), graded, and consolidated into "cull" or "shippable" racks (B). Containers shipped to the outplanting site must be returned to the nursery, which can result in damage (C), so some nurseries pack the individual cells into plastic bags (D) and cardboard boxes.

extracted as well. "Pull and wrap" is most common with large block containers, such as the Styroblock™ because extracting plants reduces the volume of space needed during storage and shipping. Harvested, dormant plants can be stored for up to 6 months, so extraction allows the used containers to be cleaned and sterilized for the next crop.

In smaller nurseries, the process of extracting plants from their containers, grading them, and packaging them is often done at individual workstations. Each station is equipped with a rack or clamp that holds the container block or tray in place while the workers pull and grade plants (fig. 7.3.8A). In larger nurseries, however, the sequence of tasks is combined into "grading and packing lines." Different workers, connected by conveyors (fig. 7.3.8B), are responsible for extracting, grading, and packaging.

Grading and packing has become more mechanized in order to reduce both labor costs and the high incidence of workplace injuries. Many forest and conservation plants have aggressive root systems and develop a firm plug by the end of the growing cycle. Roots of some species even grow into small holes in the walls of the container cavities, especially with Styrofoam™ blocks. This makes extracting plants by hand difficult, and nursery workers on the packing line often develop tendonitis and other chronic wrist and lower arm injuries. To facilitate extraction, some nurseries use mechanical "thumpers" that use a jolting motion to loosen the plugs from their containers (fig. 7.3.8C).

Another reason that plants are difficult to remove from containers is that roots often grow out of the drainage hole and form a mat (fig. 7.3.8D). To facilitate extraction, some nurseries run the container blocks over a rotating blade to sever the root mat (fig. 7.3.8E). It is much easier to prevent this by designing greenhouse benches that promote air-pruning of roots.

The extent of mechanization of the packing line varies by nursery size and sophistication. Larger nurseries often use pin or rod extractors to physically push one row of plants at a time out of the container and onto a conveyor belt, where they are graded (fig. 7.3.9A). Culls are discarded on the floor, whereas shippable plants are counted into bunches of 5 to 25. At the end of the conveyor, another worker collects the bunches and packages them.

7.3.5.3 Packaging plants

Three common packaging systems are used for native plant nursery stock.

Jellyrolling. In this first packaging system, bunches of plants are placed in a cradle and their root plugs are tightly wrapped with a protective material (fig. 7.3.10A). The jellyroll bundles are then stacked into carboard boxes for storage and shipment to the outplanting site (fig. 7.3.10B). Jellyrolling has been used for decades to protect the fine roots of bareroot stock from desiccation (Dahlgreen 1976), and research has shown that moisture stress is lower for jelly-rolled conifer stock (fig. 7.3.10C). Wet burlap and absorbent paper toweling was traditionally used, but cellophane is preferred for native plants that do not form a durable root plug. A field trial with *Ambrosia dumosa* seedlings found that jellyrolling improved their moisture status during shipping and outplanting, and that they survived and grew as well as plants shipped in containers (Fidelibus and Bainbridge 1994). The most obvious advantage of jellyrolling is that containers are not damaged or lost during the outplanting process. For polybag plants grown in native forest soil, shaking the soil from the root system, dipping the roots into a superabsorbent slurry, and jellyrolling greatly reduced the volume and weight of pine seedlings. In addition, it allowed the soil-based media to be sterilized and reused, saving the cost of procuring more media and reducing the impact on the forest environment (Mexal and others 1996).

With the recent interest in shipping frozen nursery stock to the outplanting site, jellyrolling single plants keeps them from freezing together.

Bagging and boxing. In the second packaging system, automatic bagging machines hold a supply of plastic bags that are automatically inflated by a flow of air, making it easier to insert plants (fig. 7.3.9C). In general, when bags of plants will be placed into plastic-lined boxes for storage, as is the usual case for contract crops, then the bags are just deep enough to enclose the root plugs to facilitate handling. When bundles of plants will be stored in bulk bins (fig. 7.3.11A), as is often the case for speculation crops, then the bags are large enough to enclose the entire plant, especially for evergreen crops, to retard desiccation (fig. 7.3.11B).

Figure 7.3.8—Each grading station in the "pull and wrap" operation has a rack to secure the containers (A). In larger nurseries, grading stations are part of grading and packing lines, which increase efficiency (B). Because plants are often difficult to pull from containers, such as the Styroblock™, they are first run through a "thumper" that loosens the plugs (C). If roots have formed a mat at the drainage hole (D), they must be trimmed to ease extraction (E).

Figure 7.3.9—*In more mechanized nurseries, plants are pushed from their container one row at a time by a pin extractor (A). After grading, plants are counted into bundles that are either wrapped in cellophane (B) or placed into a plastic bag (C). In the final step, bundles are placed into cardboard or plastic boxes that protect the stock during storage and shipping (D).*

Boxing. In the third packing system, which is commonly used in the southern United States, container plants that are going to be hot-planted are often extracted and placed directly into shipping boxes without any type of plastic bag (Dumroese and Barnett 2004). These plants are lifted, stored for a very short time in a cooler, and outplanted before transpiration and desiccation reduce plug moisture to an unacceptable level.

The final step in the grading and packing process involves placing the plant bundles into storage or shipping boxes or bins (fig. 7.3.9D), and marking them with the species, seedlot, number of plants, and other important information.

7.3.5.4 Processing large-volume container stock

Because of their size and weight, large container stock are typically processed one at a time and accumulated in a shadehouse or open compound until they can be shipped (fig. 7.3.12A). Although large ornamental nurseries often store their stock under refrigeration, this is not common in forestry or native plant nurseries. Square containers, such at Treepots™, are graded and stored in special metal pallet racks or in plastic crates (fig. 7.3.12B) until they can be shipped for outplanting.

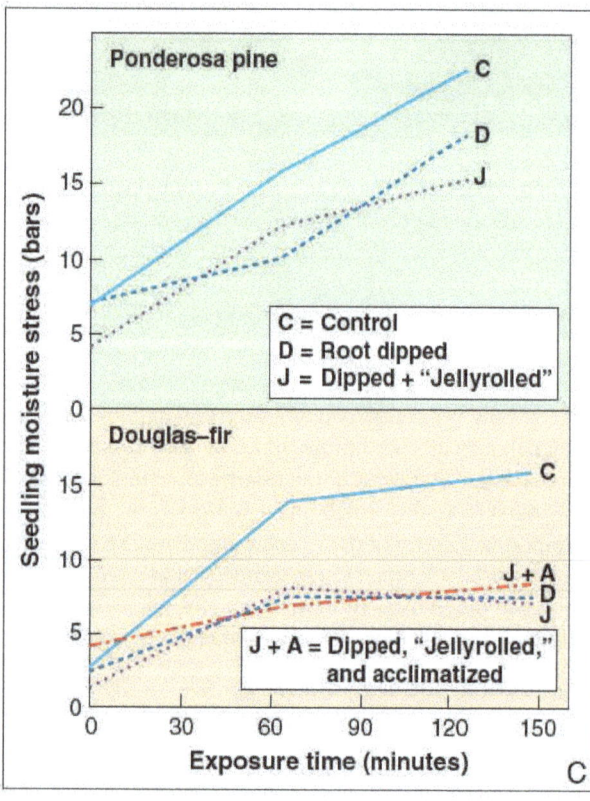

Figure 7.3.10—*Jellyrolling consists of lining up plants on a cloth, paper, or cellophane wrapper; folding the material over the roots; and then rolling the plants into a bundle (A). Research has shown that, besides protecting the root plug during storage, shipping, and outplanting (B), jellyrolling reduces plant moisture stress (C) (C, modified from Lopushinsky 1986).*

Figure 7.3.11—*When crops will be bulk-stored for later repacking and shipping (A), plants are completely inserted into deeper plastic bags to retard desiccation (B).*

Figure 7.3.12—*Large container plants are often graded in the shadehouse (A) and stored on racks until they can be shipped (B).*

7.3.6 Packaging for Storage and Shipping

Figure 7.3.13—*Waxed cardboard boxes make handling easier and storage more efficient (A). A plastic bag liner (B) is absolutely essential for long-term freezer storage because moisture is drawn out of plants and condenses on the side of the containers (C).*

The typical storage box is made of corrugated cardboard that has been treated with plastic or wax to make it waterproof (fig. 7.3.13A). Some nurseries use corrugated plastic boxes that, although they are more expensive, are reusable (fig. 7.3.13B). Even nurseries that ship stock to the outplanting site in the growth container often put stock in boxes for additional protection against mechanical injury. Boxes are the standard for refrigerated storage of pull-and-wrap container stock but, because they are not moisture-proof, a thin (1 to 2 mil) plastic bag liner is needed (fig. 7.3.13B). With freezer storage, this thin plastic bag liner is mandatory to prevent desiccation injury because refrigeration equipment continually removes excess moisture from the storage rooms (fig. 7.3.13C).

7.3.7 Processing Cull Seedlings

Figure 7.3.14—*Cull plants can be ground in a tub grinder and composted.*

At each grading station, cull plants are dropped on the floor or tossed into a bin. If there is a market, undersized but otherwise healthy plants can be transplanted into larger containers. This is common with cultivars that can be outplanted over a large geographic area or with threatened or endangered species, when every plant is valuable. Most forest and conservation plants come from a specific seed zone ("source-identified"), however, and so are only adapted to a rather restricted area. In addition, most forestry and native plant projects outplant all their stock in one season, so there is no market for holdover plants.

Therefore, most nurseries compost their culls for reuse as a soil amendment. Because the woody stems and roots would take years to decay, culls are run through a hammer mill or tub grinder to hasten decomposition and speed up the composting process (fig. 7.3.14).

7.3.8 Summary and Conclusions

Plants may be harvested, graded, packaged, and stored in a variety of ways depending on the timing of the outplanting window, the type of container used to produce the plant, and the convention used in a particular locale developed through research and/or experience. Plants harvested during the growing season with minimal dormancy and stored for just a few days with or without cooler storage are said to be "hot-planted." More often, plants are harvested when dormant and are stored for a few weeks to months in refrigerated storage. Nursery managers can determine when plants are dormant using the calendar, visual clues of the plants themselves, outplanting trials, or plant quality tests. Calculating chilling sums for a crop and correlating those with the results of plant quality tests is probably the best way for managers to ensure crops are dormant prior to harvesting.

Many factors influence the harvesting process, including nursery size and mechanization, customer base, container type, plant growth form, whether plants are extracted and subsequently placed in bags or jellyrolled, storage conditions, and local success derived from research and/or experience. For example, State and private nurseries often grow plants on speculation; those plants are extracted from containers, bundled in groups consistent with minimum orders, bulk-stored in coolers, and then packaged for shipping as orders are received. Conversely, large, reforestation nurseries usually grow plants on contract, extract them from containers, and store them in coolers or freezers until outplanting, unless those nurseries are in the Maritimes of Canada, where plants are retained in containers and stored in outdoor compounds. Many native plant nurseries grade and ship plants in their containers, particularly for species that do not produce robust root systems. As is evident, the harvesting process is dictated by many variables, but the goal of harvesting is always the same: get the crop from the nursery to the field without reducing plant quality.

7.3.9 Literature Cited

Burdett, A.N.; Simpson, D.G. 1984. Lifting, grading, packaging and storing. In: Duryea, M.L.; Landis, T.D., eds. Forest nursery manual: production of bareroot seedlings. The Hague, The Netherlands: Martinus Nijhoff Publishers: 227-234.

Burr, K.E. 2005. Personal communication. Coeur d' Alene, ID: USDA Forest Service, Coeur d' Alene nursery.

Colombo, S.J.; Gellert, S. 2002. Frost hardiness testing: an Ontario update. For. Res. Note No. 62. Sault Saint Marie, ON, Canada: Ontario Forest Research Institute. 4 p.

Colombo, S.J.; Sampson, P.H.; Templeton, C.W.G.; McDonough, T.C.; Menes, P.A.; DeYoe, D.; Grossnickle, S.C. 2001. Assessment of nursery stock quality in Ontario. In: Wagner, R.G.; Colombo, S.J., eds. Regenerating the Canadian forest: principles and practice for Ontario. Markham, ON, Canada: Fitzhenry and Whiteside: 307-323.

Dahlgreen, A.K. 1976. Care of forest tree seedlings from nursery to planting hole. In: Baumgartner, D.M.; Boyd, R.J., eds. Tree planting in the Inland Northwest. Pullman, WA: Washington State University, Cooperative Extension Service: 205-238.

Dionne, M. 2006. Personal communication. Juniper, NB: J.D. Irving, Ltd., Juniper Tree Nursery.

Dumroese, R.K.; Barnett, J.P. 2004. Container seedling handling and storage in the Southeastern States. In: Riley, L.E.; Dumroese, R.K.; Landis, T.D., tech. coords. National Proceedings: Forest and Conservation Nursery Associations—2003. Proceedings RMRS-P-33. Ogden, UT: USDA Forest Service, Rocky Mountain Research Station: 22-25.

Fidelibus, M.W.; Bainbridge, D.A. 1994. The effect of containerless transport on desert shrubs. Tree Planters' Notes 45(3): 82-85.

Fredrickson, E. 2003. Fall planting in northern California. In: Riley, L.E.; Dumroese, R.K.; Landis, T.D., tech. coords. National Proceedings: Forest and Conservation Nursery Associations—2002. Proceedings RMRS-P-28. Ogden, UT: USDA Forest Service, Rocky Mountain Research Station: 159-161.

Jackson, D.P.; Dumroese, R.K.; Barnett, J.P.; Patterson, W.B. 2007. Container longleaf pine seedling morphology in response to varying rates of nitrogen fertilization in the nursery and subsequent growth after outplanting. In: Riley, L.E.; Dumroese, R.K.; Landis, T.D., tech. coords. National Proceedings: Forest and Conservation Nursery Associations —2006. Proceedings RMRS-P-50. USDA Forest Service, Rocky Mountain Research Station: 114-119.

Jenkinson, J.L.; Nelson, J.A.; Huddleston, M.E. 1993. Improving planting stock quality—the Humboldt experience. Gen. Tech. Rep. PSW-143. USDA Forest Service, Pacific Southwest Research Station. 219 p.

Kooistra, C.M. 2004. Seedling storage and handling in western Canada. In: Riley, L.E.; Dumroese, R.K.; Landis, T.D., tech. coords. National Proceedings: Forest and Conservation Nursery Associations—2003. Proceedings RMRS-P-33. Fort Collins, CO: USDA Forest Service, Rocky Mountain Research Station: 15-21.

Landis, T.D.; McDonald, S.E. 1981. The processing, storage and shipping of container seedlings in the Western United States. In: Guldin, R.W.; Barnett, J.P., eds. Proceedings of the southern containerized forest tree seedling conference. Gen. Tech. Rep. SO-37. New Orleans, LA: USDA Forest Service, Southern Forest Experiment Station: 111-113.

Lopushinsky, W. 1986. Effect of jellyrolling and acclimatization on survival and height growth of conifer seedlings. Res. Note PNW-438. Portland, OR: USDA Forest Service, Pacific Northwest Forest and Range Experiment Station. 14 p.

Mathers, H.M. 2000. Overwintering container nursery stock, part 1: acclimation and covering. Columbus, OH: Ohio State University, Department of Horticulture, Basic Green. http://hcs.osu.edu:16080/basicgreen (accessed 4 July 2005).

Mexal, J.G.; Timmis, R.; Morris, W.G. 1979. Cold-hardiness of containerized loblolly pine seedlings: its effect on field survival and growth. Southern Journal of Applied Forestry 3(1): 15-19.

Mexal, J.G.; Phillips, R.; Landis, T.D. 1996. "Jellyrolling" may reduce media use and transportation costs of polybag-grown seedlings. Tree Planters' Notes 47(3): 105-109.

Tinus, R.W. 1996. Cold hardiness testing to time lifting and packing of container stock: a case history. Tree Planters' Notes 47(2): 62-67.

Trotter, D.; Shrimpton, G.; Dennis, J.; Ostafew, S.; Kooistra, C. 1992. Gray mould (*Botrytis cinerea*) on stored conifer seedlings: efficacy and residue levels of pre-storage fungicide sprays. In: Donnelly, F.P.; Lussenburg, H.W., eds. Proceedings: Forest Nursery Association of British Columbia meeting, 1991: 72-76.

White, B. 2004. Container handling and storage in Eastern Canada. In: Riley, L.E.; Dumroese, R.K.; Landis, T.D., tech. coords. National Proceedings: Forest and Conservation Nursery Associations—2003. Proceedings RMRS-P-33. Fort Collins, CO: USDA Forest Service, Rocky Mountain Research Station: 10-14.

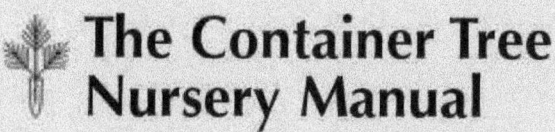

The Container Tree Nursery Manual

Volume Seven

Chapter 4
Plant Storage

Contents

7.4.1 Introduction *107*
7.4.1.1 Distance between nursery and outplanting site *107*
7.4.1.2 Differences between the lifting window at the nursery and outplanting windows *107*
7.4.1.3 Facilitating harvesting and shipping *107*
7.4.1.4 Refrigerated storage can be a cultural tool *107*

7.4.2 Short-Term Storage for Summer or Fall Outplanting—"Hot-Planting" *108*

7.4.3 Overwinter Storage *110*
7.4.3.1 Designing and locating a storage facility *110*
    The general climate at the nursery
    Characteristics of nursery stock
    Distance to the outplanting sites
    Number and size of plants to be stored

7.4.4 Nonrefrigerated Storage Systems *112*
7.4.4.1 Open storage *112*
7.4.4.2 Structureless storage systems *112*
    White plastic sheeting
    White Styrofoam™ sheets and panels
    Plastic Bubble-Wrap™ sheeting
    Frost fabrics
    Plastic film with layer of insulating material
7.4.4.3 Storage structures *114*
    Cold frames
    Cloches and polyhouses
    Shadehouses
    Greenhouses

7.4.5 Refrigerated Storage *120*
7.4.5.1 Physiology of plants in refrigerated storage *120*
    Dormancy
    Cold hardiness
    Stress resistance
    Root growth potential (RGP)
    Stored carbohydrates
7.4.5.2 Handling, thawing, and outplanting frozen stock *124*

7.4.6 Monitoring Plant Quality in Storage *126*

7.4.7 Causes of Overwinter Damage *128*
7.4.7.1 Cold injury *128*
7.4.7.2 Desiccation *128*
7.4.7.3 Loss of dormancy *128*
7.4.7.4 Storage molds *128*
7.4.7.5 Animal damage *130*

7.4.8 Summary and Conclusions *131*

7.4.9 Literature Cited *132*

7.4.1 Introduction

Unlike some agricultural commodities that can be stored for extended periods without a decrease in quality, container nursery crops are living and have a very limited "shelf life." Therefore, well-designed storage facilities are needed at all native plant nurseries.

Plant storage was not a serious consideration back in the days when nurseries were established close to the outplanting project. This allowed plants to be dug up in the nursery one day and outplanted the next. Transportation was slow and plant handling and packaging were rather simple (fig. 7.4.1). Reflecting on those days and knowing what we now do about plant physiology, it is amazing how well many of those early plantations performed.

It is important to realize that plant storage is an operational necessity, not a physiological requirement (Landis 2000), because of the following four reasons.

7.4.1.1 Distance between nursery and outplanting site

Today, most native plant nurseries are located at great distances, often hundreds or even thousands of miles, from the outplanting sites of their customers. This is particularly true of container nurseries because, as long as the proper seed source is used, high-quality plants can be grown in greenhouses in ideal growing environments located far away. The farther the distance from the nursery to the outplanting site, however, the greater the need for storage.

7.4.1.2. Differences between the lifting window at the nursery and outplanting windows

As mentioned in the previous section, container nurseries are often located in climates different from those of their customers. In mountainous areas, this is especially true, because nurseries are typically located in valleys at low elevation that have much different climates than outplanting sites at higher elevations. Differences between lifting and outplanting windows will also depend on the season of outplanting. If customers desire summer or fall outplanting, then short-term storage is all that is necessary. Often, however, the best conditions for outplanting occur the following spring, so it is necessary to protect plants throughout winter.

Figure 7.4.1—*Early forest nurseries did not need storage facilities because seedlings were shipped and outplanted within days. Note that the workers are sitting atop the bales of nursery stock.*

7.4.1.3 Facilitating harvesting and shipping

The large numbers of plants being produced at today's nurseries means that it is physically impossible to lift, grade, process, and ship stock in a short time. Therefore, one primary benefit of storage facilities is that they help to spread out the scheduling and processing during harvesting and shipping.

7.4.1.4 Refrigerated storage can be a cultural tool

Many growers do not appreciate the fact that refrigerated storage can be used to manipulate plant physiology of a variety of plants. Cold storage temperatures can partially satisfy the chilling requirement of dormant stock, and refrigerated storage has even been shown to improve plant quality (Ritchie 1989). Class 2 Douglas-fir (*Pseudotsuga menziesii*) seedlings were found to gradually increase in quality to Class 1 while in storage (fig. 7.4.2). On the other hand, plants with atypical dormancy patterns may not benefit from refrigerated storage. Cold storing water oak (*Quercus nigra*) seedlings did not appear to prolong dormancy, increase stress resistance, or increase outplanting performance (Goodman and others 2009). A complete discussion of dormancy and other aspects of plant quality can be found in Chapter 7.2.

7.4.2 Short-Term Storage for Summer or Fall Outplanting —"Hot-Planting"

Container stock that will be outplanted during summer or fall is not completely dormant or very stress resistant so it requires special consideration. The term "hot-planting" is used to describe this type of operation, because no extended period of refrigerated storage is employed. Plants are typically held in the hardening structure, which is usually a shadehouse or open compound, until they are shipped (fig. 7.3.3A). In the Southern United States, hot-planted container stock is stored in coolers or refrigerated vans at 4 to 21 °C (40 to 70 °F) for no more than a week (Dumroese and Barnett 2004).

Recent research has shown that nondormant nursery stock can perform well when hot-planted (Helenius and others 2005). Both actively growing and cold-stored Norway spruce (*Picea abies*) container stock were planted and then subjected to increasing periods of moisture stress. The nondormant seedlings that were hot-planted had significantly more new roots growing out of the container plugs ("root egress") than the cold-stored stock for the first 2 weeks after outplanting (fig. 7.4.3B).

Hot-planting can be successful in the summer and fall when conditions are ideal on the outplanting site. This system offers a lot of flexibility because plants can be held at the nursery and shipped as they are needed. On the outplanting site, nursery stock should be stored upright and kept in the shade. Using white boxes helps to reflect light and keep in-box temperatures lower (Kiiskila 1999). Hot-planting requires close coordination between the nursery and the customer; therefore, projects are usually close to the nursery and relatively small.

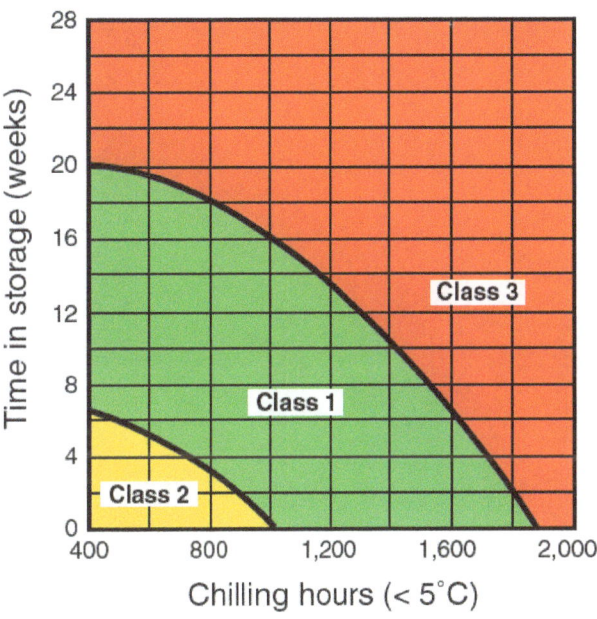

Figure 7.4.2—*Refrigerated storage can partially fulfill the chilling requirement of dormant nursery stock, and, in the case of Douglas-fir, has been shown to actually increase seedling quality—a move from the yellow to the green zone (modified from Ritchie 1989).*

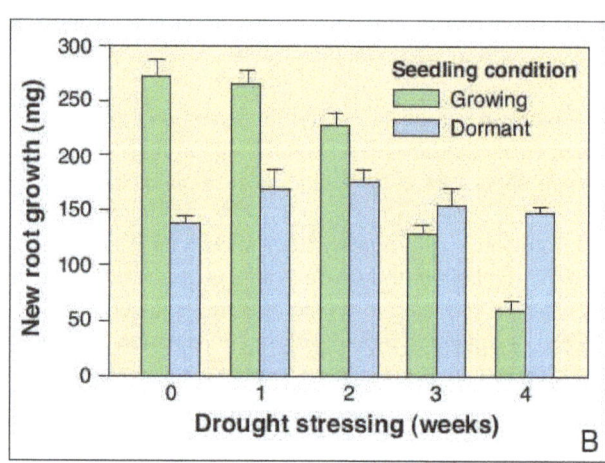

Figure 7.4.3—*Because they are not dormant or stress-resistant, hot-planted stock is held in the hardening area until shipped (A). In research trials, hot-planted spruce seedlings had better new root growth in the first 2 weeks after outplanting compared with stock that had been cold-stored (B, modified from Helenius and others 2005).*

7.4.3 Overwinter Storage

The importance of properly overwintering stock is often overlooked by novice nursery managers because they are primarily focused on growing the crop. Plants are frequently damaged and some crops have been completely lost as a result of poorly designed or managed overwinter storage (fig. 7.4.4A). Although dead plants are dramatic, what is more insidious is sublethal injury, in which roots are seriously damaged (fig. 7.4.4B). Unfortunately, sublethally damaged plants often do not develop injury symptoms under ideal nursery conditions; instead, the injury is reflected in poor survival and growth on the outplanting site. The risk for overwinter injury is very much dependent on the physiological condition of the plants at the time of storage (discussed in Chapter 7.3) and on proper storage techniques and conditions.

7.4.3.1 Designing and locating a storage facility

The time to first think about plant storage facilities is during the nursery development phase, but, unfortunately this is often not done. The design and location of a nursery storage system depends on the following four factors.

The general climate at the nursery. Most people think that overwinter storage would be more difficult the farther north or higher in elevation one goes, but that is not always the case. Nurseries in the Midwest or southern Great Plains regions of the United States are often the most challenging because of extreme weather fluctuations during winter (Davis 1994). An extreme case is the eastern slope of the Cascade Mountains or Rocky Mountains, where temperatures can vary by as much as 22 °C (40 °F) or more within a 24-hour period, and high, drying winds are common during winter and early spring. At one nursery in Alberta, a 5-year study on plant quality in outdoor compounds documented recurring damage and mortality because of late frosts and unusually warm periods during late winter (Dymock 1988). It can also be difficult to store container stock in areas such as the Southwestern United States, where winter is characterized by many clear, sunny days. Therefore, each nursery must develop a storage system appropriate for the local climate.

Characteristics of nursery stock. Some plants are easier to store than others, so storage systems must match the plant species being grown. Species that tend to overwinter well are those that achieve deep dormancy and can withstand low or fluctuating temperatures. Deciduous plants have a definite advantage because their lack of foliage when dormant reduces the possibility of winter desiccation. Evergreen species are prone to both cold injury and moisture loss, and broadleaf evergreen species are particularly troublesome. Species and ecotypes from coastal areas that are never exposed to freezing tend to be less hardy than those from inland areas. This makes it particularly challenging for nurseries that grow seedlots from a wide range of elevations. For example, coastal sources of Douglas-fir tend to grow late in the season and are much less hardy than seedlots from higher elevations in the mountains. In tropical or semitropical climates, plants never undergo true dormancy and can be outplanted almost any time of the year.

All temperate and arctic plants go through an annual cycle of growth and dormancy (see Chapter 7.2). In nurseries, plants are cultured through an accelerated period of growth that must be terminated before they

Figure 7.4.4—*Entire crops of nonhardy plants have been lost to sudden freezing temperatures when improperly stored (A). Sublethal injuries, such as cold injury to roots (B), are of greater concern because foliar symptoms are slow to develop under nonstressful nursery conditions.*

can be outplanted; this is the hardening period. In Volume 6, we discussed ways in which growers can harden their stock and prepare them for storage. Plants that are fully dormant and cold hardy are in the ideal physiological state for overwinter storage. Dormant, hardy plants can be thought of as being in a state of "suspended animation." They are still respiring and some cell division occurs in the roots and stems (see Figure 7.2.35 in Chapter 7.2); evergreen species can even photosynthesize during favorable periods during winter. The challenge to nursery managers is to design and manage a storage system to keep their stored plants dormant while protecting them from stress.

Distance to the outplanting sites. Nurseries located close to the outplanting sites may be able to hot-plant their stock with little or no storage. As the distance increases, however, some type of storage facility is needed. Nurseries that are in a climate different from their customers' climate need the most sophisticated and expensive storage systems. Because they grow stock from many different elevations with differing outplanting windows, Weyerhaeuser nurseries in Oregon and Washington use freezer storage where plants can be held for as long as 6 months (Hee 1987). The Forest Service J.H. Stone Nursery in southern Oregon has grown commercial conifers for clients across the Northwestern United States, but those from high elevations in Idaho require special handling and, therefore, incur more costs than those for local customers. On the other hand, clients in the coastal forests of Oregon can outplant throughout the winter, and so receive their plants with minimal storage.

Number and size of plants to be stored. As already mentioned, larger nurseries face a greater challenge in processing their stock, and storage systems help provide a buffer. In addition, larger container stock requires special storage considerations. For example, it is relatively easy to store a large number of 66 cm^3 (4 in^3) plants under refrigeration, but the same number of 328 cm^3 (20 in^3) plants would require four times as much space. Very large stock, such as 20-L (5-gal) containers, requires too much refrigerated storage space and so must be stored by other means.

7.4.4 Nonrefrigerated Storage Systems

Individual native plant species have distinct requirements for overwinter storage. Because of this and unique local climates, four different types of overwinter storage are commonly used in forest and conservation nurseries. Most nurseries typically use several types of storage. Three of the types of overwinter storage avoid refrigeration and are discussed in this section; refrigeration, the fourth type of storage, is discussed in Section 7.4.5.

7.4.4.1 Open storage

Open storage is the least expensive but most risky overwintering option in areas with freezing temperatures. This is especially true for small-volume container stock that has less thermal mass of the growing media to protect sensitive roots from freezing. In addition to having more thermal mass, larger containers also contain more moisture that protects against overwinter drying. Thus, the smaller the container, the higher the risk of injury.

The best locations in a nursery for open storage have some protection from the wind and are where water and cold air will drain away. Gravel and/or drainage tile should be used to promote free drainage of rain or snow melt in the spring. Packing containers together tightly on the ground and insulating their perimeter with straw bales or a berm of sawdust makes use of heat stored in the ground to protect the roots of the stored plants (fig. 7.4.5A). A research trial in Sweden showed the importance of grouping container plants and placing them directly on the ground (Lindstrom 1986). Temperatures in the peripheral containers were consistently lower than those in the interior by as much as 3 °C (5.4 °F) and fluctuated greatly. At the end of the overwinter period, plants were placed in a growth chamber to observe their performance; those stored directly on the ground had much more shoot and root growth than those stored on pallets that were 10 cm (4 in) above the ground (fig. 7.4.5B). To keep plant roots from growing into the ground, nurseries can underlay all open-stored stock with heavy poly sheeting or a copper-treated fabric that chemically prunes the roots (fig. 7.4.5C).

Open storage is most successful in forested northern climates, where adjacent trees create both shade and a windbreak, and continuous snow cover can be expected. If tree cover is not available, plants can be stored in narrow east-to-west oriented bays between vertical snowfences (fig. 7.4.5D). Snow is an ideal natural insulation for overwintering container plants, but complete and continuous snow cover is not always reliable. Some northern nurseries have had success with generating snow cover with snowmaking equipment (Davis 1994) (fig. 7.4.5E).

7.4.4.2 Structureless storage systems

Next to open storage, structureless storage systems are the simplest and least expensive ways to overwinter container stock. The term "structureless" means that plants are enclosed in a protective covering that lacks substantial mechanical support. Many different coverings have been used but the basic principle is the same—to provide a protective, insulating layer over the stored plants. Clear plastic should never be used because it transmits sunlight so that temperatures within the storage area can reach damaging levels or cause stock to lose dormancy. All plastic coverings will be eventually photodegraded by direct sunlight, so they should be stored in a dry, dark location when not in use (Green and Fuchigami 1985). Any structureless storage system is effective only if applied after plants have developed sufficient hardiness and, most important, removed before plants lose dormancy in the spring.

White plastic sheeting. Single layer films, such as a 4-mil white copolymer plastic sheeting, are the most common coverings in structureless systems. White is preferred because it reflects sunlight and keeps temperatures from buildingup under the covering. Some growers group blocks of containers together with the roots to the inside and then cover them with white plastic (fig. 7.4.6). This is less effective, however, than grouping the containers together on the ground to take advantage of the heat stored in the soil (fig. 7.4.5A).

White Styrofoam™ sheets and panels. Microfoam® is a breathable Styrofoam-like material that is lightweight, reusable, and easily removed and stored. It is available in rolls or sheets of various widths, lengths, and thicknesses (fig. 7.4.7A). Sheets can be placed directly over plants (fig. 7.4.7B) or supported by wood stakes or wiring. Because Microfoam® is so lightweight, it needs to be secured well so that it does not rip or blow away during windstorms. Typically, the foam sheets are secured around the edges

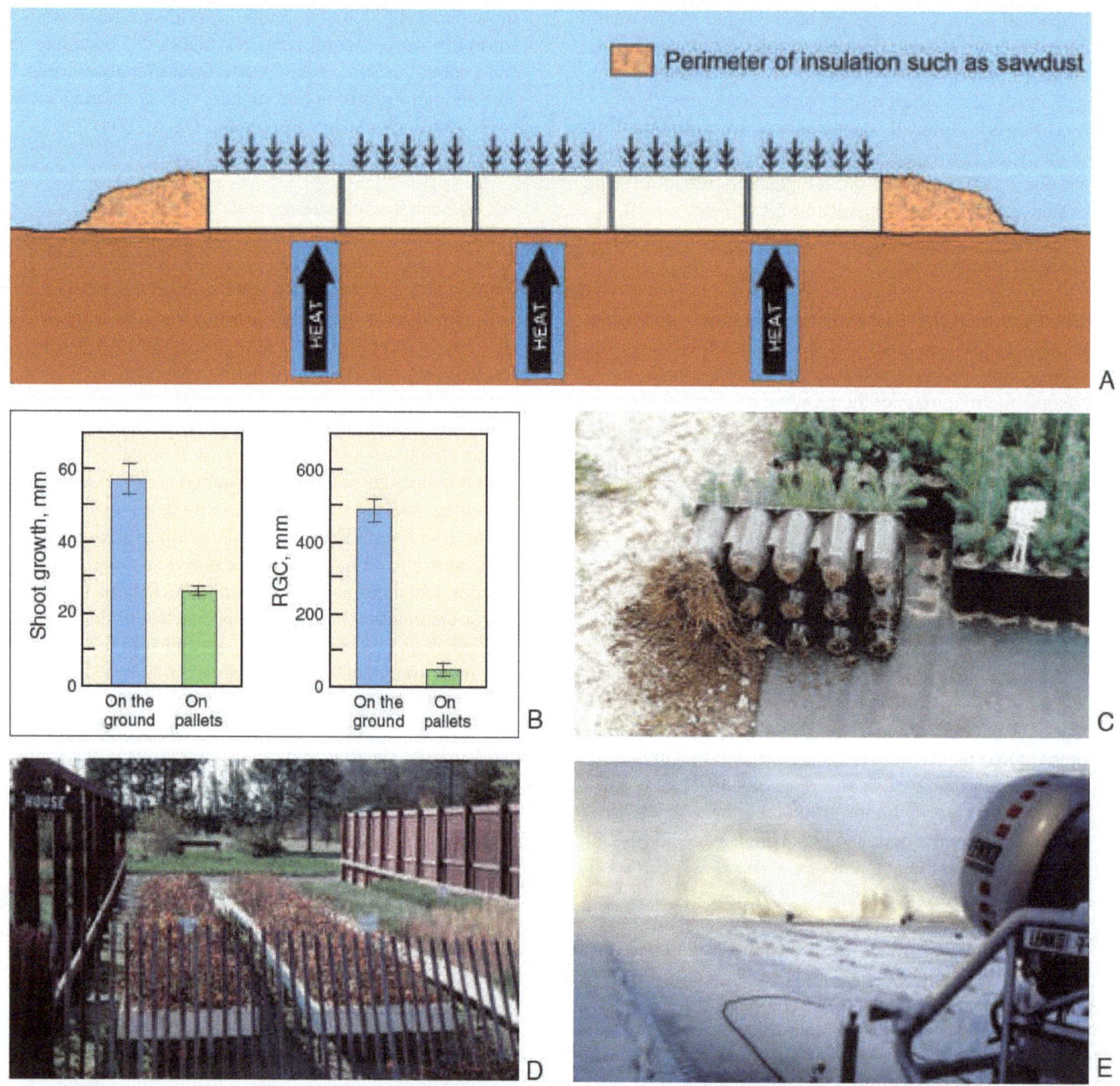

Figure 7.4.5—*Open storage can be effective when plants are blocked on the ground and surrounded by insulation (A). Both shoot growth and root growth potential of seedlings overwintered on the ground were much greater than those stored on pallets (B). Copper-treated fabrics, like Tex-R® (C), are ideal for ground storage because they chemically prevent plant roots from growing into the ground. Open-stored plants should be protected from direct sun and wind by natural or artificial snowfences (D). Snow is an excellent insulator and northern nurseries have augmented natural snowfall with snowmaking equipment (E) (B, modified from Lindstrom 1986; C, courtesy of Stuewe & Sons, Inc.; E, courtesy of Maurice Dionne).*

Figure 7.4.6—*White plastic reflects the warming rays of the sun but, by itself, has no insulation value; so it is better to leave the containers on the ground.*

with concrete blocks, wooden planks, or even a berm of sand. In a comprehensive trial in Ontario, Styrofoam SM™ blankets protected conifer seedlings from temperatures below –30 °C (–22 °F) with a significant cost savings compared to refrigerated storage (fig. 7.4.7C). The authors recommended removing insulating covers during warmer weather to allow condensation to escape and prevent overheating of stock nearest the ground. Subsequent outplanting trials produced almost identical results in survival and growth (fig. 7.4.7D) (Whaley and Buse 1994). In another test, however, one layer of Microfoam®, in the absence of reliable snow, did not provide enough protection in the harsh climates of northern Minnesota and North Dakota (Mathers 2004). As with all new techniques, nurseries considering using insulating covers should install small trials before attempting operational use.

Plastic Bubble-Wrap™ sheeting. This material has better insulation than regular plastic sheeting and is reported to be cheaper and more durable than Microfoam® sheets (Barnes 1990). Because it is clear, however, heat buildup would still be a problem on sunny days.

Frost fabrics. Woven and nonwoven landscape fabrics have been used for structureless storage. White frost fabrics retard solar heating while permitting infiltration of rain or snow melt; they allow stored plants to "breathe." Horticultural suppliers offer frost fabrics in a range of weights and thicknesses, giving 2 to 4.5 °C (4 to 8 °F) of thermal insulation. Arbor Pro® is a feltlike material that has been used successfully for conifer storage in eastern Canada (White 2004).

Plastic film with layer of insulating material. In harsh, northern climates without reliable snow cover, some nurseries cover their container stock with a "sandwich" of straw or other insulating material between two layers of clear plastic sheeting. Because the clear plastic and straw absorb solar heat on clear, frigid days, and the straw provides insulation during the night, this layering provides more overwinter protection than other structureless systems (Mathers 2003). Although layered coverings provide good insulation, they cannot be removed or vented during periods of sunny warm winter weather (Iles and others 1993).

For nurseries considering overwintering with coverings or in poly tunnels, Green and Fuchigami (1985) provide operational costs for various systems.

7.4.4.3 Storage structures

The next level of sophistication and cost is storage structures, which range from traditional cold frames to full controlled units.

Cold frames. The term "cold frame" is a traditional name for a propagation structure that receives its heat only through absorbed sunlight. When sheltered from direct sunlight and insulated, however, cold frames can be a low-cost alternative for overwinter storage. In northern Alberta and Alaska, cold frames constructed of wooden sideboards lined and topped with rigid Styrofoam™ panels have proven effective for overwintering conifer seedlings (fig. 7.4.8A). Use of insulated cold frames has resulted in a significant increase in plant survival at the Weldwood Nursery in Alberta (Matwie 1991). Cold frames constructed of wooden pallets supported by cement blocks and covered by white plastic polysheeting are considered the most cost effective overwintering system for conifer seedlings at a nursery in Eastern Canada (White 2004).

Cold frames take advantage of the heat stored in the ground, and the insulating covering retards heat loss and,

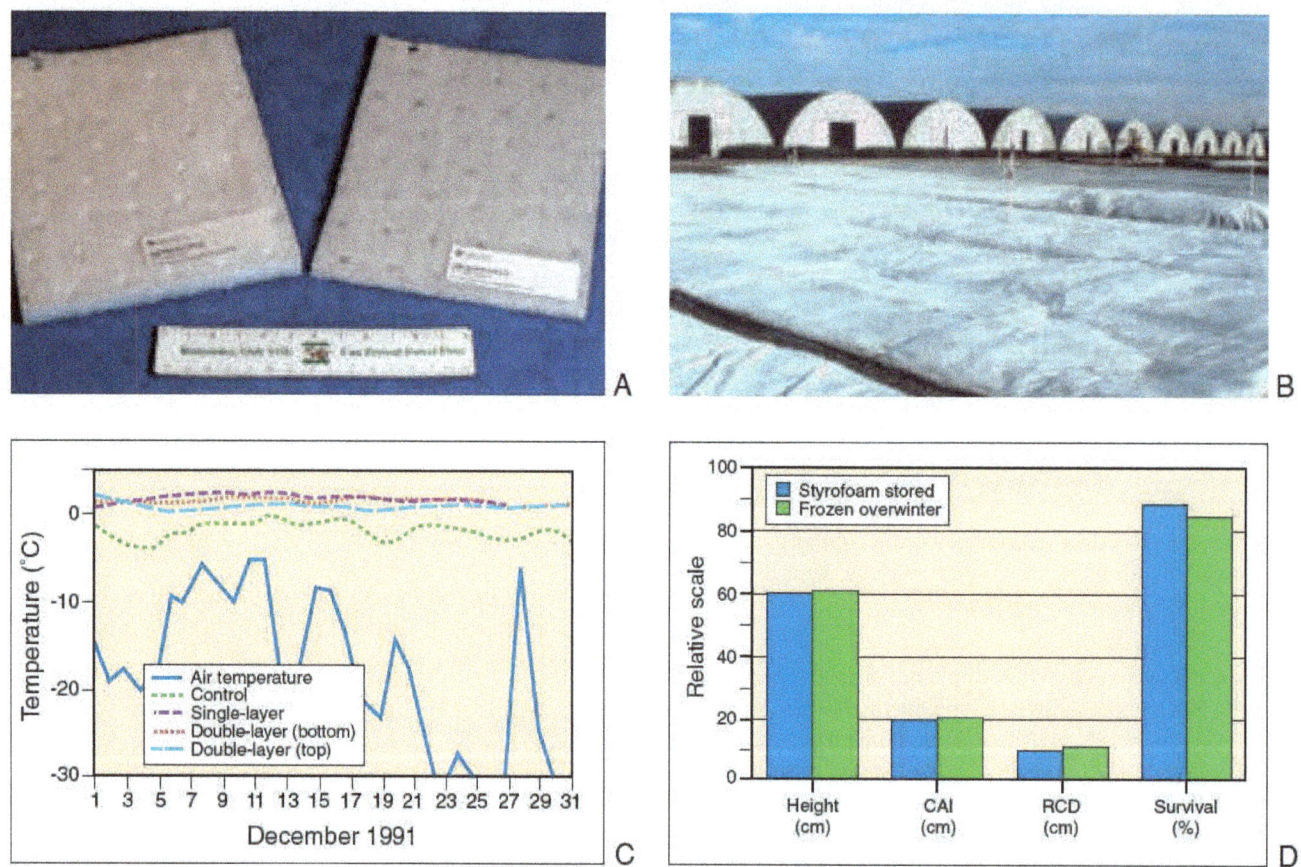

Figure 7.4.7—Microfoam® plastic foam sheeting makes an excellent overwinter cover (A). Many ornamental nurseries group their containers together on the ground and cover them with Microfoam® (B). When properly designed and applied, Styrofoam™ blankets protected conifer stock as well as refrigerated storage (C&D) (B, courtesy of Richard Regan; C&D, modifed from Whaley and Buse 1994).

115

Figure 7.4.8—*Cold frames of wood and rigid Styrofoam™ sheet insulation have been used to overwinter container plants in northern climates (A). When weather conditions permit, the top layer of insulation is removed so that plants can be irrigated (B). Cold frames can be extensive (C) and automated to protect plants during freezing temperatures (D) or retracted during heavy snowfall (E) (A&B, courtesy of Larry Matwie; C,D&E, courtesy of J.D. Irving, Limited).*

more important, prevents winter drying. To be most effective, plants should be placed in the cold frames as soon as they are hardy and before the ground freezes. Heat buildup can still be a problem during warm or sunny periods in the winter and, on these occasions, the top insulation panel can be removed for ventilation and to allow irrigation (fig. 7.4.8B). As soon as weather conditions permit in the spring, the tops of the cold frames should be removed to prevent heat buildup and subsequent loss of bud dormancy.

The Juniper Tree Nursery in New Brunswick uses large, sophisticated cold frames to overwinter their stock (fig. 7.4.8C). The accordion-like covers can be extended to protect plants from freezing temperatures or drying winds (fig. 7.4.8D), or opened during heavy snowfall (fig. 7.4.8E). Although expensive to construct, they are much less expensive than refrigerated storage (Brown 2007).

Cloches and polyhouses. These two storage structures are similar except for their length; cloches are shorter and do not offer worker access, whereas polyhouses typically have doors in the end walls. Both feature wooden or pipe frames covered by white plastic sheeting (fig. 7.4.9A) or a panel of Microfoam® placed between two layers of plastic (fig. 7.4.9B). The ends of these structures are opened for cooling during sunny, warm periods during winter (fig. 7.4.9C). Although a single layer of white polysheeting is adequate protection in milder climates, a double layer of white plastic that is inflated by a small fan provides better insulation in colder locations. In locations with frigid temperatures below –18 °C (0 °F), plants overwintered in polyhouses need the additional protection of a white polyfilm or Microfoam® blanket (Perry 1990). In milder climates, growers supply just enough heat in their polyhouses to keep the ambient temperatures just above freezing; this approach has proven effective for overwintering a wide variety of native plants in Colorado (Mandel 2004).

If possible, cloches and polyhouses should be oriented south-to-north to minimize and equalize solar heating. In structures oriented east-to-west, plants on the south side receive more light and heat than those on the north and may require irrigation during the winter. Any closed storage structure needs to be monitored carefully throughout the winter to determine if ventilation is needed on sunny days during late winter and early spring (fig. 7.4.9C).

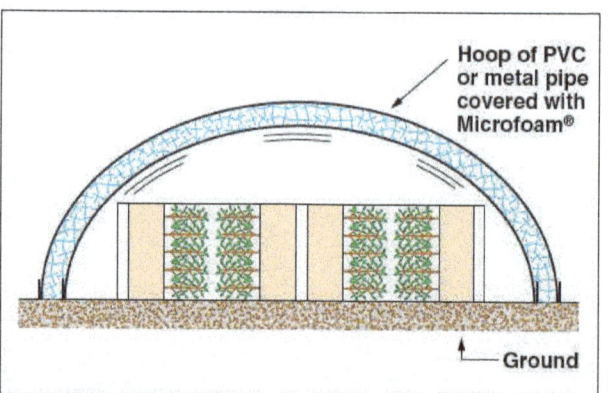

Figure 7.4.9—*Cloches and polyhouses are simple overwintering structures covered with white plastic (A) or Microfoam® sheeting (B). The ends or sides are opened during warm and sunny winter weather for ventilation (C).*

Figure 7.4.10—*Shadehouses are traditional structures that can be used for hardening and then overwintering container stock (A). They are particularly useful for large stocktypes that must be supported by heavy wire racks (B). Before freezing temperatures are expected, the containers should be grouped together on the ground and surrounded by perimeter insulation to protect the roots (C).*

Ventilation can also be provided by opening end doors, or installing a thermostatically controlled fan on one end with intake louvers at the other end. To prevent desiccation, mount fans and louvers in the top of the structures where heat buildup will be greatest.

Shadehouses. Shadehouses are traditional hardening structures that have also been used for overwintering all sizes of container stock (fig. 7.4.10A). They are particularly useful for larger container stock that requires too much space in refrigerated storage. Tall containers like Treepots™ need to be supported, so nurseries have developed heavy wire rack systems. Some nurseries use cement blocks to support prefabricated "stock panels" that can be purchased from ranch or farm supply stores (fig. 7.4.10B).

Shadehouse design varies with nursery climate and location. Where prolonged cold temperatures are not typical, plants can be overwintered under shadecloth or shadeframe. In wet climates, a waterproof roof is desirable for overwinter storage to prevent excessive leaching of nutrients from containers. In areas that receive heavy, wet snowfall, shadehouses for overwinter storage must be significantly stronger than temporary storage structures. Another option is to remove the shade covering during winter to allow the snow to fall through and insulate the crop. Light, dry snow will not damage plants and actually serves as an excellent insulator over the crop.

The typical shadehouse for overwinter storage has shading on both the roof and sides that protects plants from adverse weather, such as high winds, intense rains, hail, and heavy snow. Shadehouse storage reduces seedling temperature below what it would be in direct sunlight by reducing sunlight by about 30 to 50 percent. This shade and the reduced wind speeds significantly lower transpirational water losses; this prevents the scorching known as winter desiccation. To protect sensitive roots, plants are blocked together on the ground and surrounded by an insulating material like sawdust or Styrofoam™ panels (fig. 7.4.10C).

Greenhouses. Very sensitive plants, such as newly rooted cuttings, can be overwintered in a greenhouse with minimal heating to keep air temperatures above freezing. It must be emphasized, however, that greenhouses should

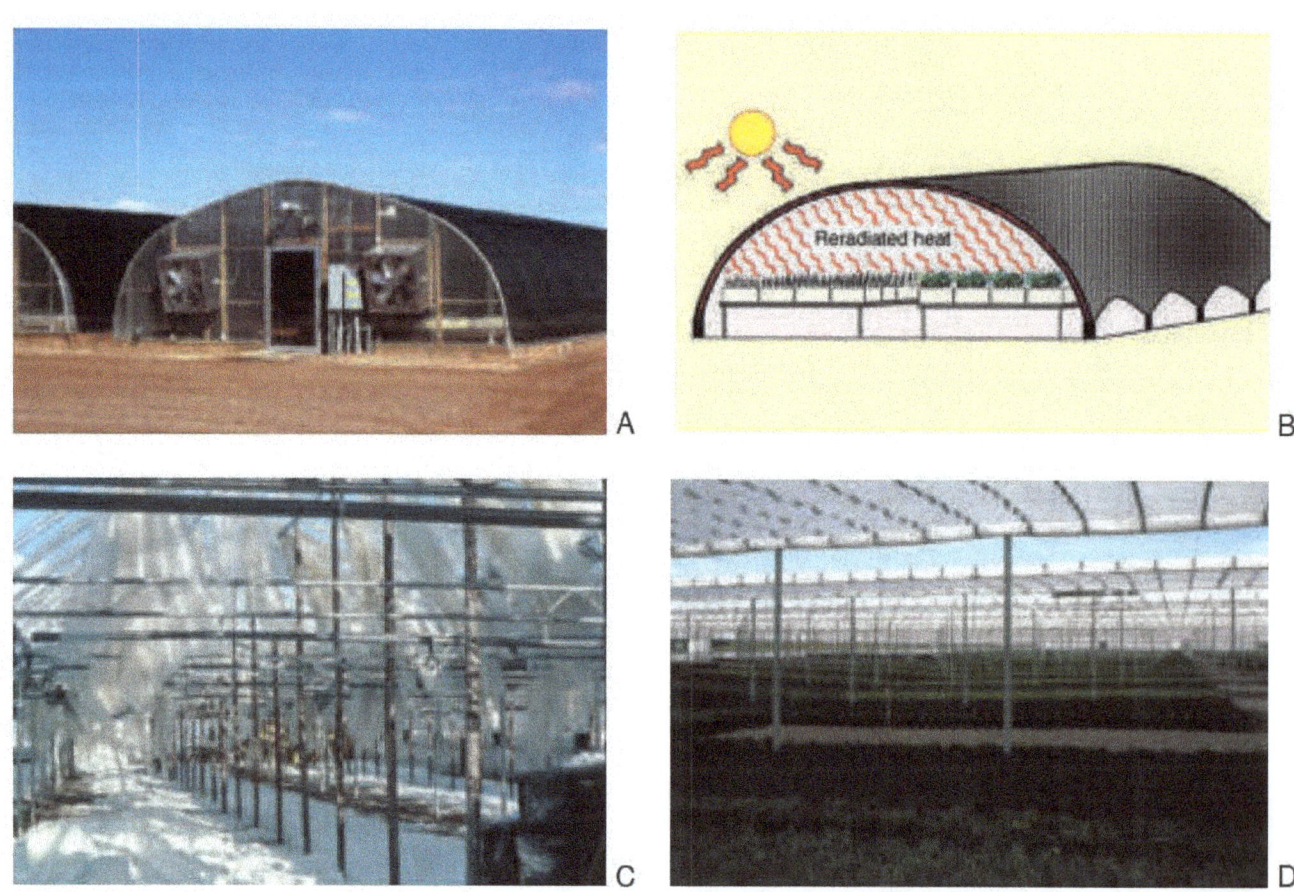

Figure 7.4.11—*Fully enclosed greenhouses are not good for overwintering, especially in climates with sunny winters (A and B). Snow removal is necessary in cold climates (C). Retractable roof greenhouses (D) are better for overwinter storage because they can be opened to allow heat to escape and snow to cover seedlings.*

not be considered for routine overwinter storage, especially in areas with clear sunny winters (fig. 7.4.11A). Greenhouses heatup rapidly during periods of sunny weather, causing plants to rapidly lose dormancy (fig. 7.4.11B). Even if greenhouses are vented, there will be considerable temperature gradients during cold weather. In snowy climates, heat must be used to keep heavy wet snow from building up and damaging the structure (fig. 7.4.11C). On the other hand, retractable roof greenhouses (fig. 7.4.11D) are excellent for overwinter storage because the roof can be opened during sunny weather to allow heat to escape and keep nursery stock dormant. During snowfall, the roof can be left open to allow plants to be covered with a protective layer of snow.

7.4.5 Refrigerated Storage

The basic concepts of refrigeration and design of refrigerated storage are covered in Section 1.3.5.4 of Volume I of this series, so this section will focus on its operational use in forest and native plant nurseries. Refrigerated storage has become the standard in many modern forestry nurseries, especially in the Pacific Northwest, and has been the focus of most storage research.

The two different types of refrigerated storage used in native plant nurseries are cooler storage and freezer storage; they are differentiated by their temperatures (fig. 7.4.12A) and the recommended duration of storage (table 7.4.1). When the photosynthetic recovery of cooler- and freezer-stored plants after outplanting was monitored, differences were minimal (fig. 7.4.12B). A review of nursery research and operational experience shows that cooler storage is best for periods of 2 months or less, whereas freezer storage is recommended for longer storage periods. Cooler storage is preferred when nursery stock is outplanted throughout the winter. For example, in the Southern States, cooler storage periods vary from a week or less in the late summer or fall to as long as 3 months (Dumroese and Barnett 2004). Although no research has been published on the subject, operational experience has shown that many broadleaved trees and shrubs store better in coolers (Davis 1994) (fig. 7.4.12B), and many other native plant species can be stored this way as well (table 7.4.2). Some species, such as black walnut (*Juglans nigra*) and dogwood (*Cornus* spp.), have serious problems with root rot in cooler storage. Considerable variation exists between species, however, so there is no substitute for practical experience.

Freezer storage has become the standard operating procedure for many commercial conifer nurseries (Hee 1987; Kooistra 2004), but less is known about how other native plants tolerate it. Because plant carbohydrate reserves decrease during cooler storage, freezer storage is recommended for storage durations longer than 2 months; even so, 6 to 8 months appears to be the practical limit for freezer storage (Ritchie 2004). Although carbohydrate reserves are conserved better with freezer storage, the primary reason for choosing freezer storage is the reduced incidence of storage molds. Because freezing converts all the free water in the storage container to ice, the development of pathogenic fungi such as gray mold (*Botrytis cinerea*) is retarded (Trotter and others 1992). After packing, plants should be frozen as quickly as possible to minimize carbohydrate loss and reduce the possibility of mold development (Kooistra 2004).

To ensure good air circulation in the storage unit, boxes of nursery stock are loaded onto pallets and then stacked onto shelving (fig. 7.4.13A) to improve air flow and prevent heat buildup. Refrigerated vans ("reefers") are sometimes used for temporary storage (fig. 7.4.13A) but are prone to breakdown and therefore are no substitute for well-designed refrigeration units.

7.4.5.1 Physiology of plants in refrigerated storage

Although refrigeration is the most expensive way to store nursery stock, it offers significant physiological advantages over other methods. As shown earlier (fig. 7.4.2), refrigerated storage can even increase the quality of stored plants. Camm and others (1994) present an excellent overview of the subject, although the authors do not always distinguish between bareroot and container stock. Ritchie (1987) is also informative. More information on all aspects of seedling quality can be found in Chapter 7.2.

Table 7.4.1—*Comparison of types of refrigerated storage*

Type of storage	In-container temperature	Recommended length of storage	Best type of packaging
Cooler storage	1 to 2 °C (33 to 36 °F)	2 weeks to 2 months	Kraft-polybags or cardboard boxes with plastic bag liners
Freezer storage	−2 to −4 °C (30 to 25 °F)	2 to 8 months	Cardboard boxes with plastic bag liners

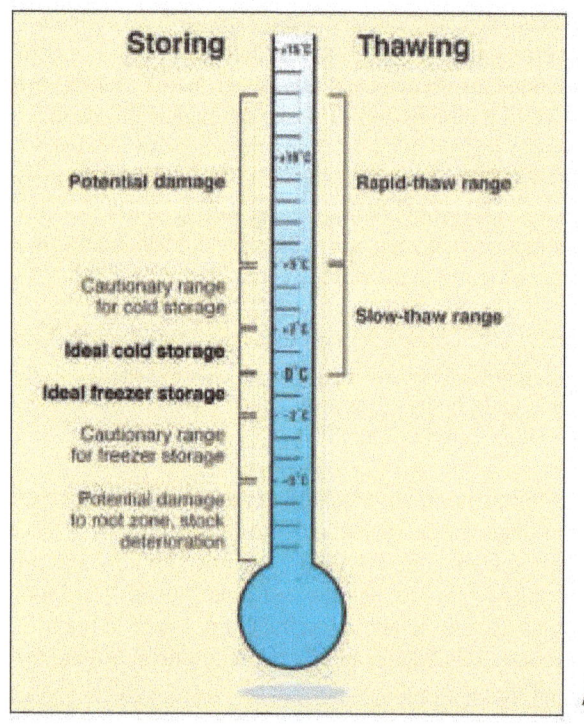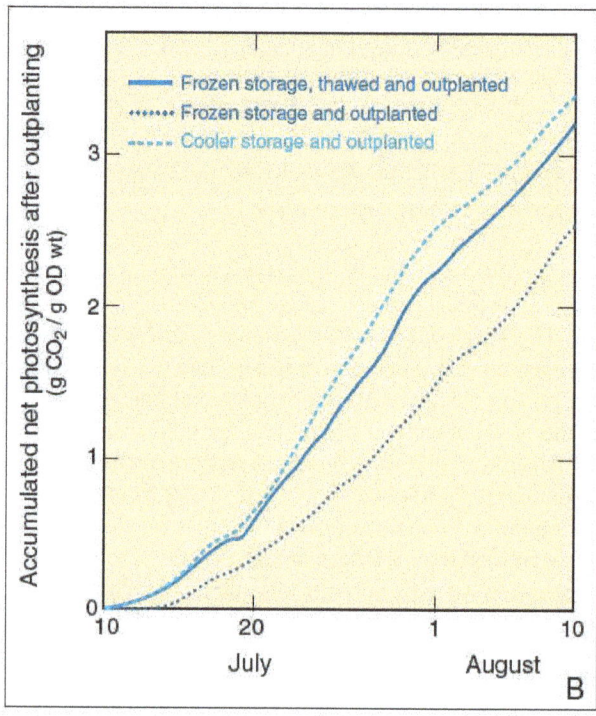

Figure 7.4.12—*The actual temperature difference between cooler and freezer storage is minimal (A), and studies of photosynthetic recovery after outplanting show little difference (B). Species differences exist, however, and operational experience has shown that some broadleaf trees and shrubs do better in cooler storage (C) (A, modifed from Paterson and others 2001; B, modified from Mattsson and Troeng 1986).*

Table 7.4.2—*Nontimber native plant storage at Coeur d' Alene Nursery varies with species and outplanting window (Burr 2004)*

Scientific name	Common name	Packaging	Type of storage	Outplanting window
Alnus rubra	Red alder	Pull, bag, and box	Freezer	Spring outplant
Alnus sinuata	Sitka alder	Pull, bag, and box	Freezer	Spring outplant
Amelanchier alnifolia	Serviceberry	Pull, bag, and box	Freezer	Spring outplant
Arctostaphylos uva-ursi	Kinnikinnick	Pull, bag, and box	Cooler	Spring outplant
Arctostaphylos uva-ursi	Kinnikinnick	Overwinter in greenhouse	Cooler	Spring regrowth, summer/fall outplant
Ceanothus velutinus	Snowbrush	Overwinter in greenhouse	Cooler	Spring regrowth, summer/fall outplant
Menziesia ferruginea	Fool's huckleberry	Overwinter in greenhouse	Cooler	Spring regrowth, summer/fall outplant
Rosa woodsii	Woods' rose	Pull, bag, and box	Freezer	Spring outplant
Rosa woodsii	Woods' rose	Pull, bag, and box	Cooler	Spring outplant
Rosa woodsii	Woods' rose	Overwinter in greenhouse	Cooler	Spring regrowth, summer/fall outplant
Salix spp.	Willows	Pull, bag, and box	Freezer	Spring outplant
Spirea betufolia	White spirea	Pull, bag, and box	Cooler	Spring outplant
Spirea douglasii	Rose spirea	Pull, bag, and box	Cooler	Spring outplant
Symphoricarpus albus	Snowberry	Pull, bag, and box	Freezer	Spring outplant
Xerophyllum tenax	Beargrass	Overwinter in greenhouse	Cooler	Spring regrowth, summer/fall outplant

Dormancy. Most research has been done on bud dormancy, and its intensity is measured by days to bud break (DBB). Refrigerated storage temperatures can partially satisfy the chilling requirement of dormant stock (Burr and Tinus 1988) and then prolong the release of dormancy until later in the spring (Dunsworth 1988). Several studies have proven that freezer storage is as effective as cooler storage in releasing dormancy (fig. 7.4.14A), provided plants have reached a certain level of cold hardiness prior to storage. In one study with white spruce (*Picea glauca*) (Harper and others 1989), dormancy release in freezer storage continued for as long as 6 months. Cooler storage and freezer storage maintain dormancy equally, and plant performance after outplanting appears to be similar. For example, when the photosynthetic recovery of Scots pine (*Pinus sylvestris*) seedlings stored in cooler or freezer storage was measured during the first season after outplanting, little difference was noted in outplanting performance (Mattsson and Troeng 1986).

Cold hardiness. Obviously, nursery stock must be cold hardy to tolerate overwinter storage, but the operational importance of cold hardiness is its relationship to overall stress resistance. Cold hardiness tests are routinely used as a storability index (see Chapter 7.2). Exactly how refrigerated storage affects the development or maintenance of cold hardiness is an important question but, unfortunately, little research has been published on container plants. In one test, interior spruce seedlings in freezer storage initially gained more cold hardiness but then lost up to half their hardiness by the end of the storage period (fig. 7.4.14B).

Stress resistance. This quality attribute reflects a plant's overall tolerance to the many physical and physiological stresses during harvesting, storage, shipping, and outplanting. Again, little research has been done with container stock, but Douglas-fir seedlings stored under refrigeration showed improved tolerance to low temperatures, root desiccation, and handling stresses (Ritchie 1986).

Root growth potential (RGP). Most studies of new root growth of plants under refrigerated storage show variable results with no discernable trends. For example, when white spruce seedlings were removed from freezer storage at intervals through the winter and were potted to observe new root growth, RGP was found to increase for 3 to 4 months and then to decrease (fig. 7.4.14C) (Harper and others 1989). This agrees with Mattsson and Lasheikki (1998) who found that RGP of Siberian larch (*Larix sibirica*) container stock decreased after about 4 months of refrigerated storage.

Stored carbohydrates. After plants are harvested and placed in dark, refrigerated storage, they begin to use stored carbohydrates, even in freezer storage (fig. 7.4.14D). Carbohydrate reserves are measured as total nonstructural carbohydrates (TNC) as opposed to structural carbohydrates, which cannot be used for energy. Ritchie (2004) estimated that conifer seedlings contain from 15 to 20 percent dry weight of TNC when they are harvested, and they gradually decrease during refrigerated storage. Obviously, the longer plants are stored under refrigeration, the less energy reserves they have for survival and growth after outplanting. Because of species differences and the wide variation of outplanting site conditions, the lower limit for TNC varies significantly. Coastal sources of Douglas-fir seedlings are in a critical area when they reach 10 to 12 percent of total dry weight (Ritchie 2004).

A

B

Figure 7.4.13—*To ensure uniform temperatures throughout the refrigerated storage units, boxes must be spaced on shelving to allow good air flow (A). Portable reefer vans (B) should be used only for short-term refrigerated storage.*

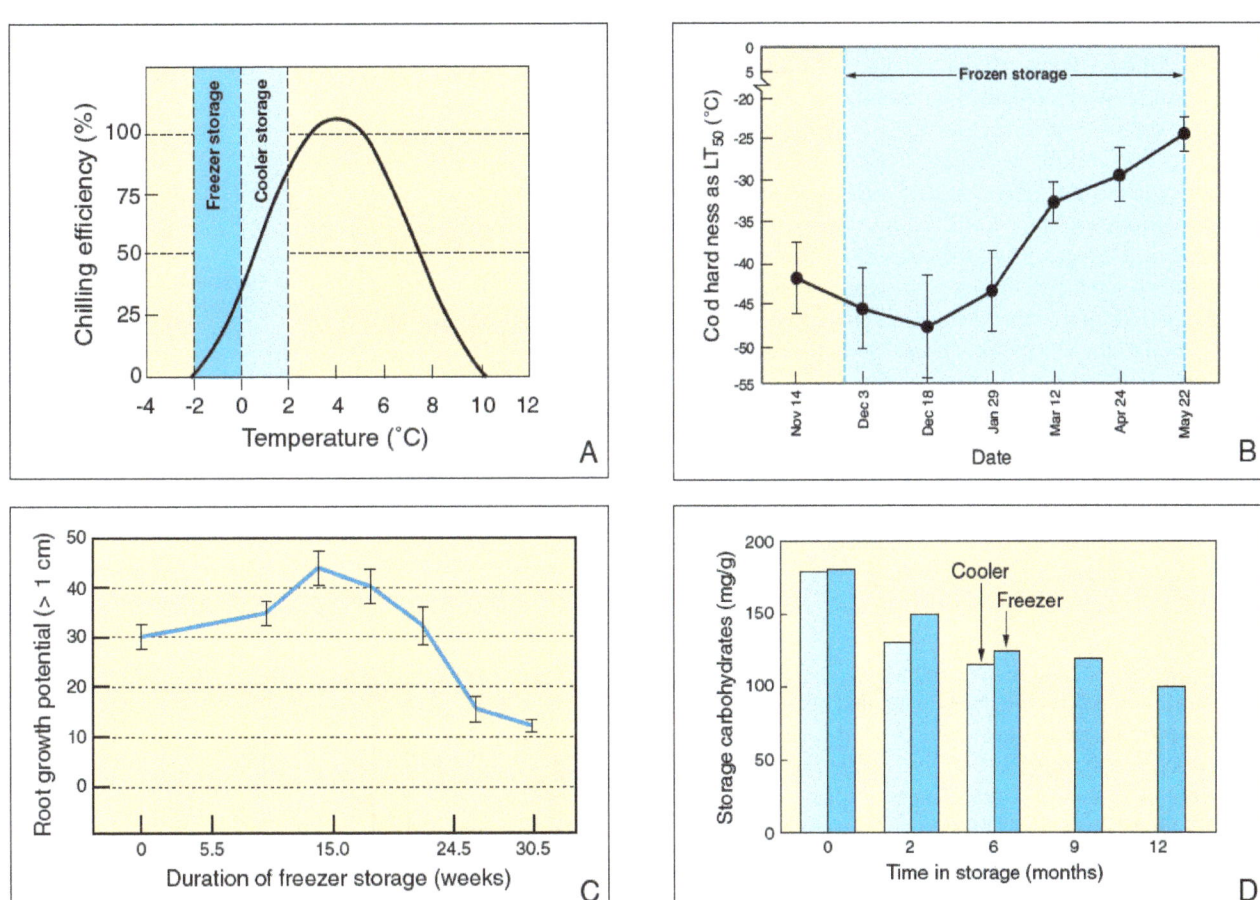

Figure 7.4.14—*Cooler and freezer storage are effective in fulfilling the chilling requirement after plants have attained a certain level of cold hardiness (A) and also have other effects on plant physiology. Compared with outside storage, conifer seedlings maintained their cold hardiness better under refrigeration (B). Root growth potential increased for about 4 months, and then declined (C). Freezer storage slows the decline in stored carbohydrates more than cooler storage and so is preferable for long-term storage (D) (A, from Ritchie 2004; B, modified from Grossnickle and others 1994; C, modified from Harper and others 1989; D, modified from Ritchie 1982).*

7.4.5.2 Handling, thawing, and outplanting frozen stock

For many nursery clients, freezer storage is a relatively new practice and some clients have expressed concern about whether frozen stock can be safely transported. Experience with commercial conifers has shown that frozen seedlings can be shipped without serious injury (Kiiskila 1999), but, like all stock, frozen stock should always be handled with care.

The speed at which frozen plants are thawed has also caused concern with many nursery customers. Initially, slowly thawing frozen stock was considered best, but rapid thawing is now gaining support. In the most comprehensive experiments to date, Camm and others (1995) studied the physiological effects of thawing regimes on the physiology of container spruce seedlings. They found no significant differences between rapid (1 to 2 days at 15 °C [60 °F]) and slow thawing (17 days at 5 °C [41 °F]). For instance, seedling moisture stress was found to recover in only 4 to 5 hours during rapid thawing (fig. 7.4.15A). These positive results were confirmed by operational trials in British Columbia (Silim and Guy 1998) that showed rapid thawing of frozen stock (15 °C [60 °F] for 1 to 2 days) resulted in less carbohydrate loss and produced better outplanting performance (fig. 7.4 15B). The Nursery Technology Cooperative at Oregon State University did a similar study and found no significant difference between slow or rapid thawing periods or for stock that was rapid thawed and then held in cold storage (Rose and Haase 1997). In one of the most well-designed and long-term studies, freezer-stored Norway spruce container stock was thawed in cardboard boxes at 39 or 54 °F (4 or 12 °C) for up to 16 days before outplanting. When outplanting survival was measured after 3 years, the best thawing temperature was 12 °C (54 °F) for 4 to 8 days, which also prevented mold development (Helenius and others 2005). Based on this research, rapidly thawing frozen stock for several days at (50 to 60 °F) can be recommended.

Obviously, common sense must be applied and thawing should be done out of direct sunlight, but it seems that the quicker the thaw, the better. The problem of increased susceptibility to storage molds is a rather specious concern, because fungal growth will be slow in cold storage.

Changing weather can shut down outplanting projects very quickly, which raises the question about what to do with the thawed plants. No research has been published on this problem, but Ritchie (2004) recommends cooler storage if the delay will be only a few days but freezing the stock if several weeks will elapse. The very latest research involves the direct outplanting of still-frozen seedlings. Comparison outplantings showed that when frozen container stock was planted, it thawed rapidly without any significant effects on plant growth (Kooistra and Bakker 2002, Islam and others 2008). This can pose an operational challenge, however, because freezer storage usually results in plants freezing together to form one large mass. Therefore, direct outplanting of frozen stock cannot be accomplished unless plants can be easily singulated.

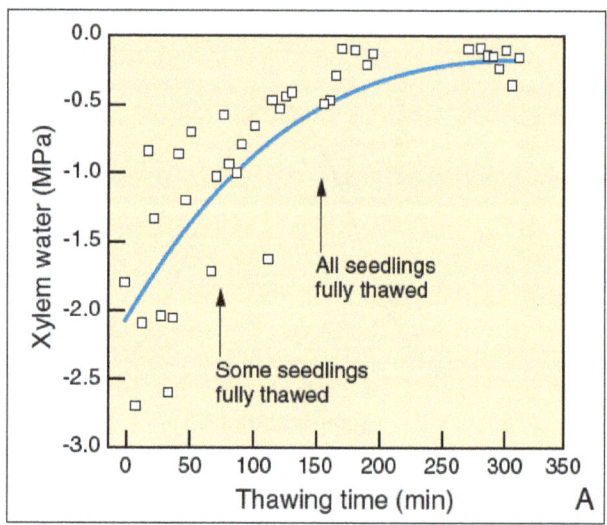

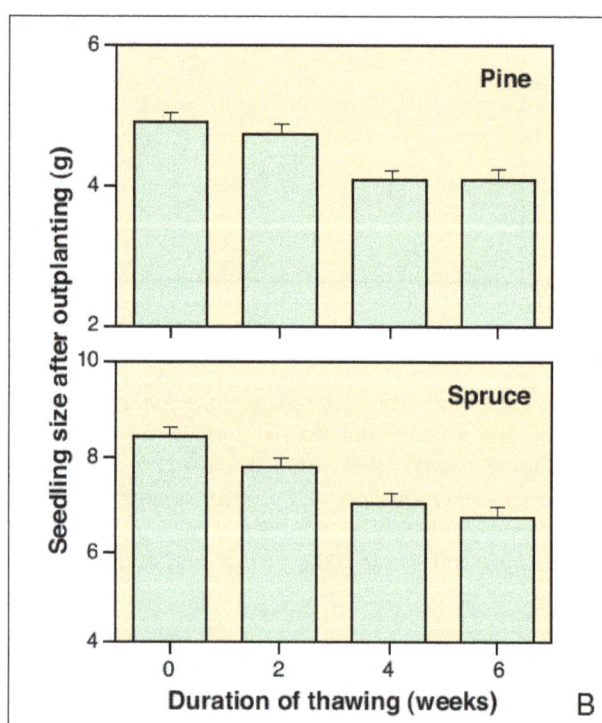

Figure 7.4.15—*Although slow thawing of frozen nursery stock was initially favored, rapid thawing has been proven to have no adverse affects in terms of plant moisture stress (A) or other physiological variables. Outplanting trials have confirmed that rapid thawing is actually beneficial to seed-ling growth (B) (A, modified from Camm and others 1995; B, modifed from Silim and Guy 1998).*

7.4.6 Monitoring Plant Quality in Storage

During overwinter storage, plants can be considered as being in a state of "suspended animation"—the plants are alive but their physiological functions have slowed to a minimum. The critical limiting factor that maintains dormancy during storage is temperature. Therefore, temperature should be rigorously monitored throughout the overwinter storage period (Kooistra 2004). Electronic thermometers with long probes are very useful for monitoring temperature in storage containers (fig. 7.4.16A). Small and inexpensive data loggers are self-contained recording devices that monitor temperature, humidity, and other weather variables that contribute to plant stress (fig. 7.4.16B). New models like the Hobo® are small enough to place in storage packages where they detect both incidence and duration of exposure (McCraw 1999). Thermochron iButtons® are even smaller and almost indestructable (Gasvoda and others 2003). Both can monitor temperature over time and the data can be downloaded to a computer (fig. 7.4.16C). Any thermometer or temperature recording device must be calibrated annually to make sure it is accurate; an easy way to do this is to place the temperature probe in a mixture of ice and water and the temperature should read exactly 0 °C (32 °F) (fig. 7.4.16D).

It is important to measure temperature within the containers as well as in the storage building, because the two locations tell different things. Because stored plants are still respiring, they generate a small amount of heat, which means that the in-bag or in-box temperature will always be a couple of degrees warmer than the ambient environment. For this reason, the setpoint temperature for the storage environment should always be 1 to 2 degrees cooler than the desired temperature in the container (Kooistra 2004). For example, you may have to operate with a setpoint of –2 °C (28 °F) to obtain a temperature of –1 °C (30 °F) in-box temperature. The temperature in the storage facility should be monitored as well, because it tells whether the compressors are working properly and good distribution of cold air is occurring (Landis 2000).

After temperature, the next most critical factor to monitor is moisture. Even hardy, dormant plants can dry out during overwinter storage. This is more of a concern with evergreen species because they will begin to transpire whenever exposed to heat and light. As mentioned earlier, desiccation is a continual threat in refrigerated storage, especially freezer storage. Even deciduous species can be damaged, so it is important to check plants occasionally during the storage period and to irrigate, if necessary. Because water is so heavy, weighing storage boxes is the most accurate way to monitor moisture loss during storage. Many nurseries measure plant moisture stress with a pressure chamber prior to and during harvesting (see Chapter 7.2); this equipment also affords a quick and accurate way for monitoring the degree of plant desiccation during storage (Landis 2000).

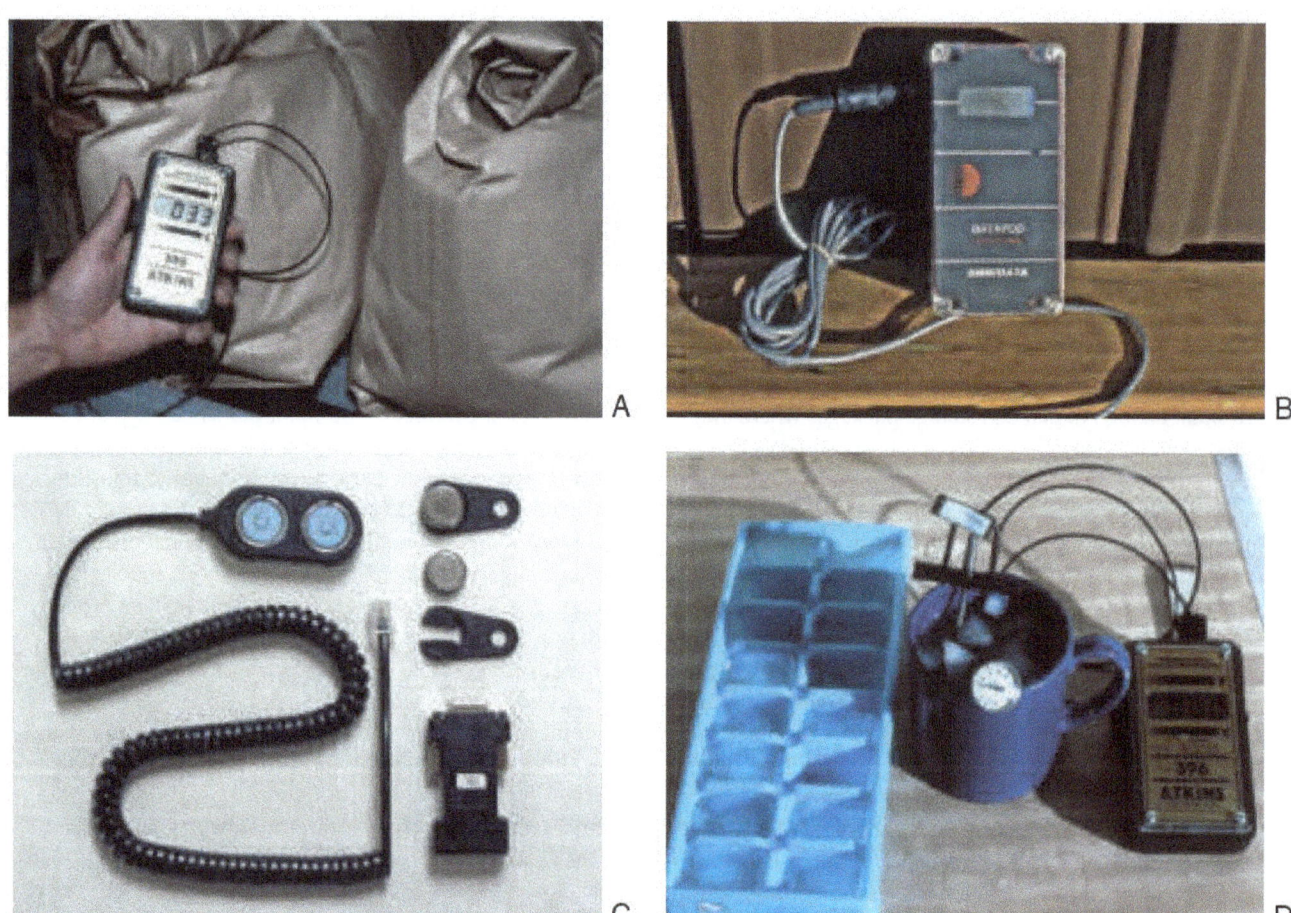

Figure 7.4.16—*Temperature can be monitored with long-stemmed electronic thermometers (A). Small hygrothermographs like the Datapod® can monitor both temperature and relative humidity (B). Even smaller, the iButton® can monitor temperature for weeks or months, and the data can be downloaded to a computer (C). Calibrate any thermometer in a water and ice mixture to make sure it is accurate (D).*

7.4.7 Causes of Overwinter Damage

Overwinter storage has many potential hazards for stored plants (table 7.4.3), and growers should periodically monitor stock for the following hazards.

7.4.7.1 Cold injury

Cold injury can develop from a single frost or during an extended period of cold weather. Damage is most common in the late fall or early spring, when plants are entering or coming out of dormancy. Cold injury is directly related to plant dormancy or cold hardiness. The shoots of native plants that have been properly hardened can tolerate freezing temperature extremes expected in the geographic area from which they originate, but cold hardiness and dormancy are lost as winter progresses. The lateral meristem at the root collar can be injured by frost (fig. 7.4.17A), as can the buds. This type of injury is very difficult to diagnose without destructive sampling, because symptoms may not become evident until later in the spring.

Root systems need special protection because they are injured at much higher temperatures than shoots. Furthermore, young fibrous roots are less hardy than older woody roots and will be injured at higher temperatures. Rooted cuttings are particularly vulnerable to injury because their roots have not yet developed protective layers. Young roots are typically on the outside of the container and are the first to be injured by cold temperatures (fig. 7.4.17B). Where freezing temperatures occur, cold injury to roots is the most common type of overwinter damage. Because shoots do not show symptoms immediately, root injury often goes unnoticed and the damage becomes evident after outplanting. Therefore, growers should design their overwintering systems to protect all roots from damaging temperatures during overwinter storage.

7.4.7.2 Desiccation

Winter drying is actually desiccation injury and occurs whenever plants are exposed to extreme moisture stress, especially wind and/or direct sunlight (fig. 7.4.17C). Damage is most severe when the growing medium and roots remain frozen for extended periods while the shoots are exposed to sun and wind. Plants can even become desiccated when they are stored under frost-free refrigeration without proper packaging. Winter drying is not directly related to plant dormancy or cold hardiness—even the most dormant and hardy stock can be damaged. Plants stored near the perimeter of open compounds or sheltered storage are most susceptible (fig. 7.4.17D), but even plants covered by snow can be damaged if their tops become exposed. This type of desiccation can be prevented if open or shelter-stored stock can be irrigated during the winter storage period and if effective perimeter insulation is used.

7.4.7.3 Loss of dormancy

Loss of dormancy happens most often when container stock is overwintered in greenhouses. During winter periods of clear, sunny weather, greenhouses can heat up and cause plants to lose dormancy. Loss of dormancy becomes progressively more serious during late winter and early spring, when plants have fulfilled their chilling requirements and cold temperatures are the only thing keeping them from growing (fig. 7.4.17E). Although refrigerated storage is the best prevention, using white or reflective coverings in structureless storage minimizes the effects of sunlight and prevents heat buildup. Monitor the temperature frequently in sheltered storage and ventilate if necessary.

7.4.7.4 Storage molds

The type of storage conditions will determine the types of disease problems that may be encountered. Although fungal diseases can be a problem in open storage or shadehouses, they are most serious when plants are overwintered under refrigeration (table 7.4.3). Some fungi, such as *Botrytis cinerea*, actually prefer the cold, dark conditions in storage bags and boxes and will continue to grow and damage plants whenever free moisture is available (fig. 7.4.17F). Some nurseries apply fungicides before overwinter storage, but careful grading to remove injured or infected plants is the best prevention. Freezer storage has become popular because it prevents the further development of storage molds. Much more information is provided in Section 5.1.6 in Volume Five of this series.

Figure 7.4.17—*Overwinter storage is a time of considerable risk for nursery stock. Cold temperatures can damage nonhardy tissue, such as the lateral meristem (A). Roots are particularly susceptible because they will grow whenever temperatures permit (B). Winterburn (C) is actually desiccation and is particularly severe around the perimeter of storage areas (D). Overwintered plants gradually lose dormancy and can break bud during late winter or early spring (E). Storage molds (F) are most serious in cooler storage, whereas animal damage can be a real problem in sheltered storage (G).*

7.4.7.5 Animal damage

The only type of overwinter storage where animals will not pose a threat is refrigerated storage. Small rodents, such as mice and voles, can be pests in shadehouses and structureless systems (fig. 7.4.17G), because the pests are protected from natural predators and from harsh weather conditions. Baiting or trapping to keep populations low can be effective if started early in the season. Large animals, such as deer and rabbits, can be pests in open, structureless, and shadehouse storage, but fencing is an effective way to prevent damage. See section 5.1.6 in Volume Five of this series for more specific information.

Table 7.4.3—*Plants can be injured by several types of stresses during overwinter storage*

Type of damage	Cause	Preventative measures for types of storage		
		Open	Sheltered	Refrigerated
Cold injury (fig. 7.4.17A-B)	Temperatures below plants' cold hardiness level Roots are much more susceptible than shoots	Properly harden plants to tolerate maximum expected cold temperatures		
Drought injury (Winter desiccation) (fig. 7.4.17C-D)	Exposure to intense sunlight and especially drying wind	Fully saturate media before storing		
		Shade plants and protect from wind	Cover stock with moisture-retaining film	Use moisture-retentive packaging
Loss of dormancy (fig. 7.4.17E)	Temperatures above 5 °C (40 °F)	Not possible	Monitor and ventilate as needed	Maintain cold in-box temperatures.
Storage molds (fig. 7.4.17F)	Warm temperatures; latent infections of *Botrytis*	Prevent injury to plant tissue; cull damaged plants		
		Keep foliage cool and dry	Keep foliage cool and dry	Use freezer storage if stored more than 2 months
Animal damage (fig. 7.4.17G)	Small rodents and even rabbits can girdle stored nursery stock	Exclude larger animals with fencing; use poison bait for rodents		Not a problem

7.4.8 Summary and Conclusions

Nondormant plants destined for nearby outplanting may go from the nursery to the field with little or no storage (hot-planted). More commonly, dormant plants are stored during winter until they can be outplanted. Storage becomes more important as the distance from the nursery to the outplanting sites increases, differences between climates at the nursery and field sites are great, and nurseries produce large quantities of plants requiring months to process. Therefore, storage is an operational necessity rather than a physiological requirement.

Overwinter storage should be developed to meet local climate, plant type, and production factors. In general, three types of overwinter storage are used: open, structureless, and structured. In open storage, plants are left outdoors, on the ground, and are protected from sun and wind by larger trees and snowfall. Plants stored in a structureless system are also outdoors and on the ground, but they are protected from the vagaries of winter weather by various applications of plastic and/or Styrofoam™ sheets. Structured storage can be very simple, such as a cold frame, progressing through modest structures, such as polyhouses and shadehouses that provide some climate control, to the most complex systems—refrigerated units. Refrigerated storage includes coolers (temperatures just above freezing), which are best for short-term storage of plants (2 weeks to 2 months), and freezer storage (temperatures just below freezing), which is best for long-term storage (2 to 8 months).

Regardless of the type of storage used, plants should be regularly monitored to ensure that pests (animals and storage molds) are not becoming a problem, temperatures are in the proper range to keep plants dormant, and medium moisture is appropriate to avoid desiccation.

After storage, plants should be shipped carefully to the field. Stock kept in freezer storage can be safely shipped while still frozen, but if thawed at the nursery, the thawing process should be rapid to reduce carbohydrate losses and development of storage molds.

Successful storage of container plants is one of the most challenging and important aspects of nursery management. Many types of overwintering systems can be employed, depending on location, climate, and species grown; more than one system may be used at a nursery. Determining when it is safe to harvest plants so that they will maintain a high level of quality throughout the storage period and to the outplanting site is one of the most challenging aspects of nursery management.

7.4.9 Literature Cited

Barnes, H.W. 1990. The use of bubble-pac for the overwintering of rooted cuttings. Combined Proceedings of the International Plant Propagators' Society 40: 553-557.

Brown, K.E. 2007. Personal communication. Juniper, NB: J.D. Irving, Ltd., Juniper Tree Nursery.

Burr, K.E. 2004. Personal communication. Coeur d' Alene, ID: USDA Forest Service, Coeur d' Alene nursery.

Burr, K.E.; Tinus, R.W. 1988. Effect of the timing of cold storage on cold hardiness and root growth potential of Douglas-fir. In: Landis, T.D., ed. Proceedings, combined meeting of the Western Nursery Associations. Gen. Tech. Rep. RM-167. Fort Collins, CO: USDA Forest Service, Rocky Mountain Forest and Range Experiment Station: 133-138.

Camm, E.L.; Goetze, D.C.; Silim, S.N.; Lavender, D.P. 1994. Cold storage of conifer seedlings: an update from the British Columbia perspective. Forestry Chronicle 70(3): 311-316.

Camm, E.L.; Guy, R.D.; Kubien, D.S.; Goetze, D.C.; Silim, S.N.; Burton, P.J. 1995. Physiological recovery of freezer-stored white and Engelmann spruce seedlings planted following different thawing regimes. New Forests 10(1): 55-77.

Davis, T. 1994. Mother nature knows best. Nursery Manager 10(9): 42-45.

Dumroese, R.K.; Barnett, J.P. 2004. Container seedling handling and storage in the Southeastern States. In: Riley, L.E.; Dumroese, R.K.; Landis, T.D., tech. coords. National Proceedings, Forest and Conservation Nursery Associations—2003. Proceedings RMRS-P-33. Fort Collins, CO: USDA Forest Service, Rocky Mountain Research Station: 22-25.

Dunsworth, B.G. 1988. Impact of lift date and storage on field performance for Douglas-fir and western hemlock. In: Landis, T.D., ed. Proceedings, combined meeting of the Western Nursery Associations. Gen. Tech. Rep. RM-167. Fort Collins, CO: USDA Forest Service, Rocky Mountain Forest and Range Experiment Station: 199-206.

Dymock, I.J. 1988. Monitoring viability of overwintering container stock in the Prairies–an overview of a five year lodgepole pine study. In: Landis, T.D., ed. Proceedings, combined meeting of the Western Nursery Associations. Gen. Tech. Rep. RM-167. Fort Collins, CO: USDA Forest Service, Rocky Mountain Forest and Range Experiment Station: 96-105.

Gasvoda, D.S.; Tinus, R.W.; Burr, K.E. 2003. Monitor tree seedling temperature inexpensively with the Thermochron iButton Data logger. Tree Planters' Notes 50(1): 14-17.

Goodman, R.C.; Jacobs, D.F.; Apostol, K.G.; Wilson, B.C.; Gardiner, E.S. 2009. Winter variation in physiological status of cold stored and freshly lifted semi-evergreen *Quercus nigra* seedlings. Annals of Forest Science 66(103). 8 p.

Green, J.L.; Fuchigami, L.H. 1985. Overwintering container-grown plants. Corvallis, OR: Oregon State University, Department of Horticulture. Ornamentals Northwest Newsletter 9(2): 10-23.

Grossnickle, S.C.; Major, J.E.; Folk, R.S. 1994. Interior spruce seedlings compared with emblings produced from somatic embryogenesis. I. Nursery development, fall acclimation, and over-winter storage. Canadian Journal of Forest Research 24(7): 1376-1384.

Harper, G.; Camm, E.L.; Chanway, C.; Guy, R. 1989. White spruce: the effect of long-term cold storage is partly dependent on outplanting soil temperatures. In: Landis, T.D., ed. Proceedings, Intermountain Forest Nursery Association. Gen. Tech. Rep. RM-184. Fort Collins, CO: USDA Forest Service, Rocky Mountain Forest and Range Experiment Station: 115-118.

Hee, S.M. 1987. Freezer storage practices at Weyerhaeuser nurseries. Tree Planters' Notes 38(2): 7-10.

Helenius, P.; Luoranen, J.; Rikala, R. 2005. Physiological and morphological response of dormant and growing Norway spruce container seedlings to drought after outplanting. Annals of Forest Science 62: 201-207.

Iles, J.K.; Agnew, N.H.; Taber, H.G.; Christians, N.E. 1993. Evaluations of structureless overwintering systems for container-grown herbaceous perennials. Journal of Environmental Horticulture 11: 48-55.

Islam, M.A.; Jacobs, D.F.; Apostol, K.G.; Dumroese, R.K. 2008. Transient physiological responses of planting Douglas-fir seedlings with frozen or thawed root plugs under cool-moist and warm-dry conditions. Canadian Journal of Forest Research 38: 1517-1525.

Kiiskila, S. 1999. Container stock handling. In: Gertzen, D.; van Steenis, E.; Trotter, D.; Summers, D.; tech. coords. Proceedings of the 1999 Forest Nursery Association of British Columbia. Surrey, BC, Canada: British Columbia Ministry of Forests, Extension Services: 77-80.

Kooistra, C.M. 2004. Seedling storage and handling in western Canada. In: Riley, L.E.; Dumroese, R.K.; Landis, T.D., tech. coords. National Proceedings, Forest and Conservation Nursery Associations—2003. Proceedings RMRS-P-33. Fort Collins, CO: USDA Forest Service, Rocky Mountain Research Station: 15-21.

Kooistra, C.M.; Bakker, J.D. 2002. Planting frozen conifer seedlings: warming trends and effects on seedling performance. New Forests 23(3): 225-237.

Landis, T.D. 2000. Seedling lifting and storage and how they relate to outplanting. In: Cooper, S.L., comp. Proceedings of the 21st Annual Forest Vegetation Management Conference. Redding, CA: 27-32.

Lindstrom, A. 1986. Outdoor winter storage of container stock on raised pallets—effects on root zone temperatures and seedling growth. Scandinavian Journal of Forest Research 1(1): 37-47.

Mandel, R.H. 2004. Container seedling handling and storage in the Rocky Mountain and Intermountain regions. In: Riley, L.E.; Dumroese, R.K.; Landis, T.D., tech. coords. National Proceedings, Forest and Conservation Nursery Associations—2003. Proceedings RMRS-P-33. Fort Collins, CO: USDA Forest Service, Rocky Mountain Research Station: 8-9.

Mathers, H.M. 2003. Summary of temperature stress issues in nursery containers and current methods of production. HortTechnology 13(4): 617-624.

Mathers, H.M. 2004. Personal communication. Columbus, OH: assistant professor, extension specialist: nursery and landscape. Ohio State University, Department of Crop and Soil Science.

Mattsson, A.; Lasheikki, M. 1998. Root growth in Siberian larch (*Larix sibirica* Ledeb.) seedlings seasonal variations and effects of various growing regimes, prolonged cold storage and soil temperatures. In: Box, J.E., Jr., ed. Root demographics and their efficiencies in sustainable agriculture, grasslands and forest ecosystems. Dordrecht, The Netherlands: Kluwer Academic Publishers: 77-88.

Mattsson, A.; Troeng, E. 1986. Effects of different overwinter storage regimes on shoot growth and net photosynthetic capacity in *Pinus sylvestris* seedlings. Scandinavian Journal of Forest Research 1(1): 75-84.

Matwie, L. 1991. Overwintering in insulated coldframes improves seedling survival. Unpublished report. Hinton, AB, Canada: Weldwood of Canada Ltd. 4 p.

McCraw, D. 1999. Onset Hobo temp recorder. In: Landis, T.D.; Barnett, J.P., eds. National Proceedings, Forest and Conservation Nursery Association—1998. Gen. Tech. Rep. SRS-25. Asheville, NC: USDA Forest Service, Southern Research Station: 3-4.

Paterson, J.; DeYoe, D.; Millson, S.; Galloway, R. 2001. Handling and planting of seedlings. In: Wagner, R.G.; Colombo, S.J., eds. Regenerating the Canadian forest: principles and practice for Ontario. Sault Saint Marie, ON, Canada: Ontario Ministry of Natural Resources: 325-341.

Perry, L.P. 1990. Overwintering container-grown herbaceous perennials in northern regions. Journal of Environmental Horticulture 8:135-138.

Ritchie, G.A. 1982. Carbohydrate reserves and root growth potential in Douglas-fir seedlings before and after cold storage. Canadian Journal of Forest Research 12(4): 905-912.

Ritchie, G.A. 1986. Relationships among bud dormancy status, cold hardiness, and stress resistance in 2+0 Douglas-fir. New Forests 1(1): 29-42.

Ritchie, G.A. 1987. Some effects of cold storage on seedling physiology. Tree Planters' Notes 38(2): 11-15.

Ritchie, G.A. 1989. Integrated growing schedules for achieving physiological uniformity in coniferous planting stock. Forestry (Suppl) 62: 213-226.

Ritchie, G.A. 2004. Container seedling storage and handling in the Pacific Northwest: answers to some frequently asked questions. In: Riley, L.E.; Dumroese, R.K.; Landis, T.D., tech. coords. National Proceedings, Forest and Conservation Nursery Associations—2003. Proceedings RMRS-P-33. Fort Collins, CO: USDA Forest Service, Rocky Mountain Research Station: 3-6.

Rose, R.; Haase, D.L. 1997. Thawing regimes for freezer-stored container stock. Tree Planters' Notes 48(1-2): 12-17.

Silim, S.N.; Guy, R.D. 1998. Influence of thawing duration on performance of conifer seedlings. Forest Nursery Association of British Columbia meetings, proceedings, 1995, 1996, 1997. Surrey, BC, Canada: British Columbia Ministry of Forests, Extension Services: 155-162.

Trotter, D.; Shrimpton, G.; Dennis, J.; Ostafew, S.; Kooistra, C. 1992. Gray mould (*Botrytis cinerea*) on stored conifer seedlings: efficacy and residue levels of pre-storage fungicide sprays. In: Proceedings, Forest Nursery Association of British Columbia meeting: 72-76.

Whaley, R.E.; Buse, L.J. 1994. Overwintering black spruce container stock under a Styrofoam® SM insulating blanket. Tree Planters' Notes 45(2): 47-52.

White, B. 2004. Container handling and storage in Eastern Canada. In: Riley, L.E.; Dumroese, R.K.; Landis, T.D., tech. coords. National Proceedings, Forest and Conservation Nursery Associations—2003. Proceedings RMRS-P-33. Fort Collins, CO: USDA Forest Service, Rocky Mountain Research Station: 10-14.

The Container Tree Nursery Manual

Volume Seven

Chapter 5
Handling and Shipping

Contents

7.5.1 Introduction *137*

7.5.2 Minimizing Stresses During Handling *138*
7.5.2.1 Moisture stress *138*
7.5.2.2 Temperature stress *139*
        Increased hazard from storage molds
        Accelerated resumption of growth
        Moisture stress
        Heat stress
        Freezing damage
7.5.2.3 Physical stress *140*
7.5.2.4 Accumulated stress *141*

7.5.3 Handling and Shipping Systems *142*
7.5.3.1 Shipping in the growth container *142*
7.5.3.2 Shipping in boxes or bags *142*

7.5.4 Nursery Stock Delivery *145*
7.5.4.1 Delivery in refrigerated trailers *145*
7.5.4.2 Delivery in nonrefrigerated trucks *146*
        Small pickup trucks
        Commercial parcel trucks

7.5.5 Summary and Recommendations *148*

7.5.6 Literature Cited *149*

7.5.1 Introduction

Nursery plants are in a period of high risk from the time they leave the protected environment of the nursery to when they are outplanted. Good guidelines for proper care during this critical time have been published for bareroot nursery stock (DeYoe 1986; USDA Forest Service 1989), and the same guidelines apply to container plants. During handling and shipping, nursery stock may be exposed to many damaging stresses, including extreme temperatures, desiccation, mechanical injuries, and storage molds (table 7.5.1). This is also the period of greatest financial risk, because nursery plants have reached their maximum value right before shipping (Paterson and others 2001). Adams and Patterson (2004) concluded that improper handling of nursery stock was a more important factor than the type of outplanting tool.

One reason for the increasing preference of container stock is that it tolerates the abuses of storage, shipping, and handling better than bareroot plants. This is particularly true with many broadleaved trees and other native plants; for example, oak (*Quercus* spp.) and beech (*Fagus* spp.) seedlings grown in a variety of containers tolerated rough handling better than bareroot stock (fig. 7.5.1). In a hardwood restoration outplanting, even the highest quality nursery plants did not survive and grow well if they were not handled properly (Self and others 2006).

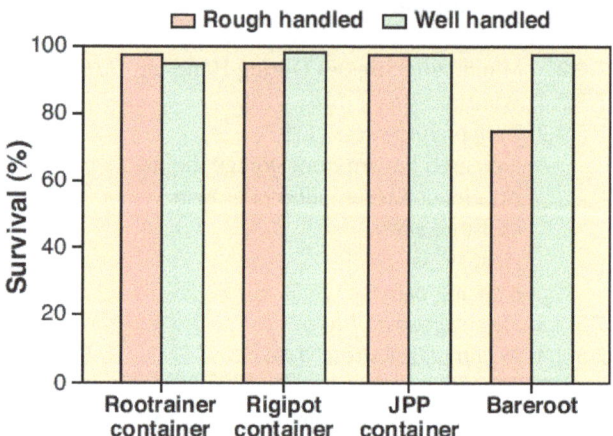

Figure 7.5.1—*Oak and beech container plants tolerated rough handling much better than did bareroot stock (Kerr 1994).*

Table 7.5.1—*Nursery plants are subjected to a series of potential stresses, from harvest through outplanting*

Process	Types of stress			
	Temperature extremes	Desiccation	Mechanical injuries	Storage molds
Nursery storage	High	Low	None	Medium
Handling	Medium	Medium	High	None
Shipping	Medium	Low	High	None
Onsite storage	High	High	None	High
Outplanting	High	High	High	None

Potential levels of stress: ▢ None ▢ Low ▢ Medium ▢ High

7.5.2 Minimizing Stresses During Handling

It is important to emphasize that nursery plants are alive and perishable, and so should be treated with utmost care at all times. Stressful injuries incurred between lifting from the nursery and outplanting, however, are often not evident until several weeks after planting. Symptoms include browning, chlorosis, poor survival, or decreased growth and are commonly known as "transplant shock" or "check." It can be extremely difficult to pinpoint the exact stress that leads to these symptoms (fig. 7.5.2A). It is a waste of time and money to produce or purchase high-quality plants only to have them die or grow poorly after outplanting as a result of these unnecessary stresses. As emphasized in Chapter 7.2, plants are best able to tolerate stress when they are not actively growing. Nonhardened, succulent plant tissue is much more vulnerable to stresses (fig. 7.5.2B). Regular monitoring of plant condition, close supervision of nursery and field personnel, periodic testing of plant quality, and maintenance of detailed records are essential to document conditions during shipping and handling.

Three stresses are most common after stock leaves the nursery: moisture, temperature, and physical.

7.5.2.1 Moisture stress

Desiccation is the most common stress encountered during handling, shipping, and storage at the field site (onsite storage) and can have a profound effect on survival and growth. Plant water potential influences every physiological process, and at stressful levels, can greatly reduce growth, even if survival is unaffected. These damaging effects can persist for several seasons after outplanting.

Roots are the most vulnerable to desiccation because, unlike leaves and needles, they have no waxy coating or stomata to protect them from water loss. Fine root tips have a greater moisture content than woody roots and are most susceptible to desiccation. If fine roots appear dry, then they are probably already damaged although it is difficult to quantify the amount of injury in the field. When exposed for just 5 minutes, bareroot conifer seedlings exhibited increasing moisture loss with increasing air temperature and wind speed (fig. 7.5.3). This shows the critical importance of keeping nursery plants cool, out of direct sunlight, and protected from drying winds.

Figure 7.5.2—*It is often difficult to determine exactly which stresses lead to "transplant shock" or "check" (A). Nonhardy and nondormant nursery stock (B) is much more susceptible to all types of stresses during handling and shipping.*

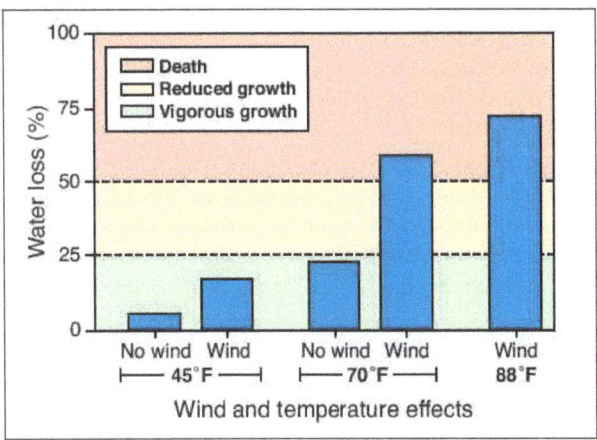

Figure 7.5.3—*When bareroot conifer nursery stock was exposed for 5 minutes, plant moisture loss increased with higher temperatures and wind until plant survival and growth were adversely affected (modified from Fancher and others 1986).*

Fortunately, roots of container plants are protected somewhat by the growing medium, which serves as a reservoir of water and nutrients. If the plug is allowed to get too dry, however, desiccation damage can be severe. Once roots have dried, subsequent growth reductions are inevitable, even when shoot water potential recovers (Balneaves and Menzies 1988). Dormant conifer plants are more vulnerable to damage from root exposure than are dormant hardwood plants.

Moisture stress can be avoided by making sure plugs are kept moist (but not saturated) throughout their journey from nursery to outplanting. Container stock should be irrigated 1 to 2 days before harvesting, depending on weather conditions (Fancher and others 1986). This allows the plugs to drain to field capacity; saturated media is unhealthy for roots, increases shipping and handling weight, and increases the potential for storage molds.

7.5.2.2 Temperature stress

Either hot or cold temperature extremes can quickly reduce the quality of nursery plants during handling and shipping.

Exposure to warm temperatures can damage stock in several ways.

Increased hazard from storage molds. Pathogenic fungi, such as Botrytis mold, can survive in all types of storage and may grow rapidly during shipping in the humid environment of a storage bag or box if the temperatures are too warm. Increased carbon dioxide from plant respiration in storage and shipping containers is also thought to stimulate fungal development. There have been anecdotal reports of storage mold "blowups" in boxes of freezer stored nursery stock after only a few days exposure to ambient conditions. Storage molds are discussed in detail in Volume Five.

Accelerated resumption of growth. Nursery plants that are stored during the winter are harvested at peak hardiness, which is ideal for storage, shipping, and handling. When ready for outplanting, properly stored plants have had their chilling requirements fully satisfied, however, and cold temperature is the only environmental factor that prevents resumption of growth. After their chilling requirement has been met, stored nursery stock exposed

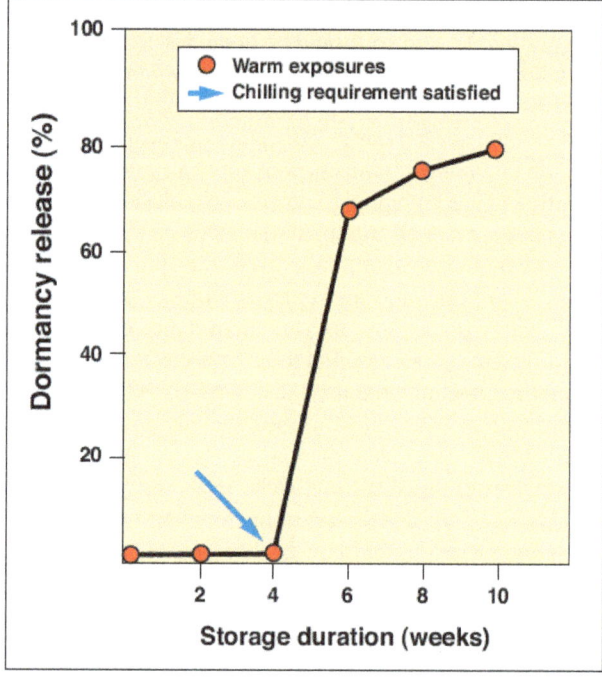

Figure 7.5.4—*Cold-stored Norway spruce (Picea abies) seedlings exposed to short periods of warm temperatures (17 °C [63 °F]) rapidly broke dormancy after the chilling requirement had been met (modified from Hanninen and Pelkonen 1989).*

to even a short period of warm temperatures will rapidly initiate shoot growth (fig. 7.5.4).

Moisture stress. Stagnant air within the storage or shipping bag or box is a poor heat conductor, but direct sunlight and wind can rapidly increase plant temperatures and cause serious moisture stress (fig. 7.5.3).

Heat stress. Stored nursery plants are alive and respiring. This means, when plants are exposed to warm temperatures, their respiration adds heat to their environment; this is particularly serious in closed environments such as storage or shipping bags or boxes. Maintaining good air circulation in storage areas, especially in nonrefrigerated storage, will minimize heat buildup due to plant respiration.

Freezing damage. Freezing temperatures can damage nursery stock. Because they are much less cold-hardy, roots are much more susceptible than shoots to freeze

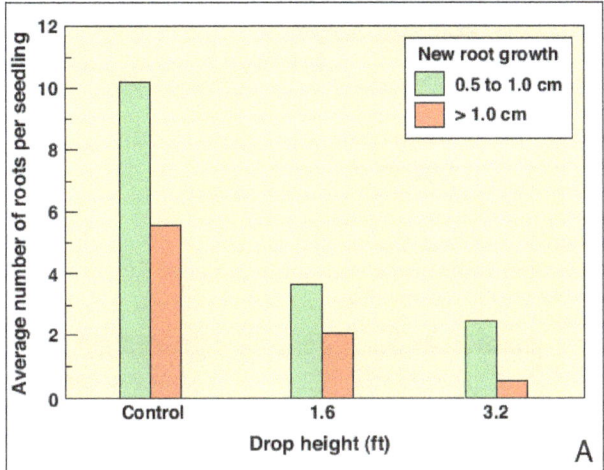

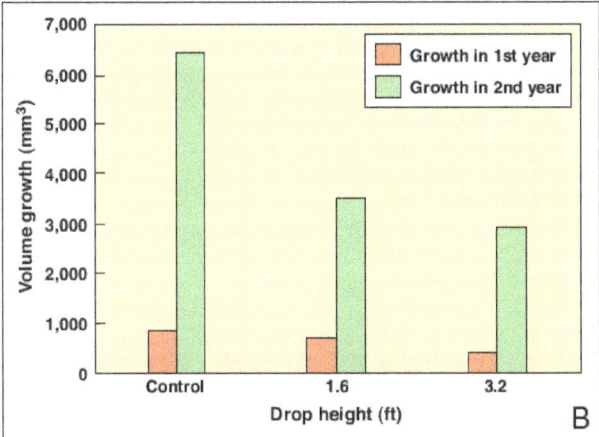

Figure 7.5.5—*When bags of conifer seedlings were dropped from different heights, their ability to produce new roots (root growth potential) was significantly reduced (A). This mechanical injury still affected plant growth 2 years after outplanting (B) (modified from Stjernberg 1996).*

damage. Ambient and in-box temperatures should be monitored regularly; temperature monitoring equipment is now inexpensive and readily available (See Section 7.4.6). Freezing damage has even occurred in cooler storage during shipping because of equipment failure. This is common, because refrigeration units on shipping vans are notoriously fickle and air circulation is restricted. Boxes in the front of the van near the refrigeration units will necessarily be colder than those in the back. Resist the temptation to overpack trucks; leave adequate space for good air circulation (Rose and Haase 2006). Stock that has been cooler stored should be shipped at these same temperatures (0.5 to 1 °C [33 to 34 °F]), whereas frozen stock can be shipped under warmer temperatures to begin the thawing process.

When nursery stock reaches the outplanting site, the plants should be kept as cool as possible. Onsite storage is discussed in Chapter 7.6.

7.5.2.3 Physical stress

Boxes of nursery plants are handled many times from when they leave the nursery until the plants are finally outplanted. Rough handling can result in reduced plant performance after outplanting. Each person involved in the hand-ling and shipping of nursery stock should receive training on how to minimize physical stresses.

The potential for physical damage to nursery stock can come from dropping, crushing, vibrating, or just rough handling. It is easy to forget that nursery plants are alive when they are in boxes. Studies have shown that the stress of dropping boxes of seedlings reduced root growth potential, decreased height growth, increased mortality, and increased fine-root electrolyte leakage (McKay and others 1993; Sharpe and others 1990; Tabbush 1986). Stjernberg (1996) did a comprehensive evaluation of the physical stresses that nursery stock is subjected to during transport from the nursery to the outplanting site. Root growth potential tests on boxes of cooler-stored white spruce seedlings showed fewer new roots were produced as the distance the box was dropped increased (fig. 7.5.5A). Interestingly enough, volume growth of these seedlings still showed growth depression 2 years after outplanting (fig. 7.5.5B).

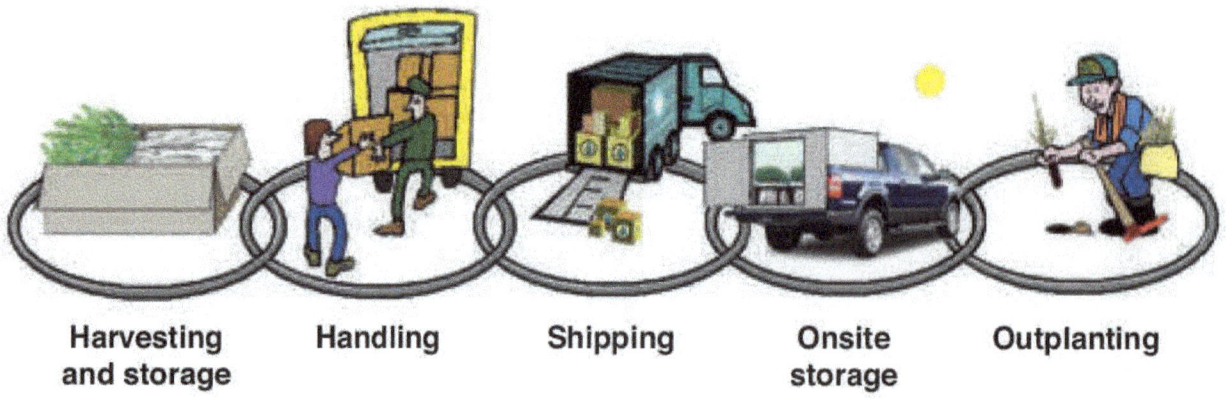

Harvesting and storage **Handling** **Shipping** **Onsite storage** **Outplanting**

Figure 7.5.6—*Nursery plants are subjected to a series of stresses from the time they are harvested to when they are outplanted. Each stage in the process represents a link in a chain, and overall plant quality is only as good as the weakest link.*

7.5.2.4 Accumulated stress

Nursery plants are at their maximum quality immediately before they are harvested at the nursery, but they then must pass through many hands before being outplanted. Outplanting success is dependent on maintaining plant quality by minimizing stress at each phase of the operation. It is useful to think of plant quality as a chain in which each link represents one of the sequences of events from harvesting and storage at the nursery until planting at the outplanting site (fig. 7.5.6). The cumulative effect of the various stresses can be much greater than any one individual stress.

As stress increases, the plant shifts energy from growth to damage repair. Physiological functions are damaged and survival and growth are reduced. These effects are exacerbated further when plants are outplanted on harsh sites.

Extremely careless handling of planting stock usually manifests itself immediately after outplanting—plants die within days or weeks. Unfortunately, the ramifications of poor handling are usually more insidious and are not immediately apparent because it causes a degree of sublethal injury that only will be reflected in decreased survival and growth weeks or months after outplanting. A good example is root injury. Roots that have been damaged by exposure or freezing may not look any different, but they have lost the ability to function properly. This condition is particularly serious with container stock because these injuries primarily affect roots on the outside of the root plug. Because the roots on the inside still function, damaged plants are able to remain turgid and so appear normal. After outplanting, however, the damaged roots cannot grow out into the surrounding soil, so plants struggle for a while and may eventually die. On moist sites with low evaporative demand, this can take weeks or months.

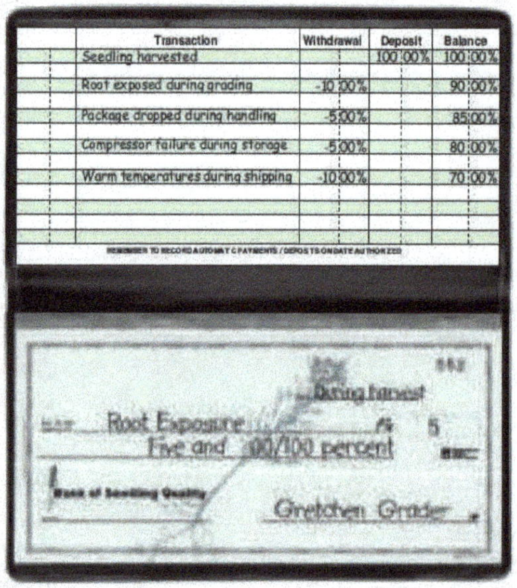

Figure 7.5.7—*It is useful to think of nursery plant quality as a checking account in which all types of abuse or stress are withdrawals. Note that all stresses are cumulative and no deposits can be made—it is impossible to increase plant quality after nursery harvest.*

Because all types of abuse or exposure are cumulative, it is helpful to think of nursery plant quality as a checking account. Immediately before harvesting, plants should be at 100 percent quality, but all subsequent stresses are withdrawals from the account (fig. 7.5.7). It is impossible to make a deposit—nothing can be done to increase plant quality after plants leave the nursery.

7.5.3 Handling and Shipping Systems

When nursery stock is ready to outplant, it must be moved from nursery storage to the outplanting site. Nursery employees usually use the same handling system to move plants in and out of storage, and the equipment is generally conveyors, handcarts, forklifts, and other motorized handling equipment that was discussed in Volume One. Shipment to the customer or outplanting site, however, often requires specialized handling and equipment. The best handling system for shipment will depend on plant physiological condition, container type and size, and whether the plants will be shipped in their containers.

7.5.3.1 Shipping in the growth container

When container nursery stock was first produced back in the 1970s, most nurseries shipped plants to their customers in the growth container. Some nurseries still use this practice and usually stack containers on metal or wooden racks inside the delivery van (fig. 7.5.8A). This method works best when shipping distances are relatively short and the roads are not overly rough. Some nurseries place their containers inside cardboard boxes for shipping. Unfortunately, used containers need to be returned to the nursery, requiring a second trip, and often the containers are damaged during transit (fig. 7.5.8B).

Shipping in containers has worked well for forest industry nurseries that grow stock for their own lands. For example, J.D. Irving, Limited, a reforestation company in New Brunswick, Canada, has developed a sophisticated pallet handling system for moving containers about the nursery and to the outplanting site (figs 7.5.8 C&D). Their system allows efficient handling of container plants while providing excellent protection against mishandling. After the stock is outplanted, the same pallets are used to transport empty containers back to the nursery without damage (Brown 2007).

One primary advantage of this system is that plant roots are protected by the container and, if plants must be held in temporary storage before planting, they can be irrigated (fig. 7.5.8E). Nursery stock that will be hot-planted is not fully hardy and endures less stress when left in the growth container. The major disadvantage of shipping in containers is that a given number of plants occupies considerably more volume and weighs more compared with stock that has been extracted for storage. In addition, stock shipped in the containers, especially those that cannot be consolidated, is not necessarily graded, so cull plants may also be handled and shipped.

Large container stock (> 500 ml [> 1 qt]) is always shipped in its growth container because it is too large and heavy to be handled any other way. The Forest Service J.H. Stone Nursery in Central Point, Oregon, grows native plants in containers up to 55 L (15 gal) in volume (fig. 7.5.9A). These plants are grown in special racks at the nursery; these same racks are used to transport the plants to the outplanting site (fig. 7.5.9B).

7.5.3.2 Shipping in boxes or bags

Nursery stock extracted from growth containers and stored in cardboard boxes requires much less storage space and weighs less when shipped as compared with stock shipped in their containers. In addition, the box provides physical protection during storage, shipping, and handling (fig. 7.5 10A). Boxes stack efficiently (fig. 7.5.10B) and pallets of boxes can be easily moved by hand, pallet jack, or forklift and loaded quickly and readily into delivery vans (fig. 7.5.10C). Delivery vans should be equipped with racks; otherwise, the weight of boxes stacked too high could mechanically damage the stock (fig. 7.5.10D).

Figure 7.5.8—*Shipping plants in their growth containers requires a rack system to support and protect the stock (A). One drawback of this system is that containers must be returned to the nursery and can be damaged in transit (B). Some large forest nurseries have developed sophisticated rack handling systems for shipping plants to the outplanting site and returning the used containers to the nursery (C&D). One advantage of shipping in the growth container is that plants can be irrigated prior to outplanting (E) (C&D, courtesy of J.D. Irving, Limited).*

Figure 7.5.9—*Large container stock is always shipped in the growth container and transported to the field in the same nursery racks (A&B).*

Figure 7.5.10—*Cardboard boxes provide protection to plants during storage, shipping, and handling (A). Because they can be stacked, boxes make efficient use of space in storage units and delivery vans (B). Pallets of boxes can be easily moved by hand or forklift (C). Delivery vans should use racks to prevent shipping boxes from being crushed (D).*

7.5.4 Nursery Stock Delivery

Many different methods have been used to deliver plants from the nursery to the outplanting site. The most appropriate method will depend on the distance involved, the number of plants, and the dormancy and hardiness of the stock. Although rail and even commercial airlines have been used, most nursery stock is delivered by truck because most outplanting sites are in remote locations. Nursery plants can be subjected to severe mechanical shocks during transport, especially on gravel or dirt roads, and reducing speed will minimize potential injury (Stjernberg 1997).

7.5.4.1 Delivery in refrigerated trailers

Nursery stock, whether shipped in the container or extracted and packed into boxes or bags, is typically shipped by truck. For large quantities of plants, typical of forest companies, and when the trip will take more than a few hours, shipping is usually done by trucks with refrigerated trailers ("reefers") (fig. 7.5.11A). Because high temperature is the major risk factor during nursery stock transport, the use of refrigerated vans has had a significant effect on enhanced plant quality and outplanting success. In a review of the success of southern pine plantations, the use of refrigerated vans was named as the single most important factor in making sure that plants arrived at the outplanting site in good condition (Fox and others 2007).

The risk of injury to nursery stock increases with the shipping distance. This is most critical for hot-planted stock for summer or fall outplanting because the plants are not fully dormant or at their peak of hardiness. When the outplanting performance of hot-planted silver birch (*Betula pendula*) seedlings was monitored, height growth measured after the first year was not affected by transportation distance (Luoranen and others 2004). When the same plants were measured after 3 years, however, shoot height decreased with shipping distance (fig. 7.5.11B).

The temperature inside the trailer should be monitored during transit because truck refrigeration units are prone to failure. Both high- and low-temperature injury has occurred when refrigerated units have malfunctioned. The ideal temperature in a delivery truck depends on whether the plants are going to be hot-planted or are coming from cooler or freezer storage, especially if frozen stock needs to thaw. In an operational trial, temperatures in boxes of

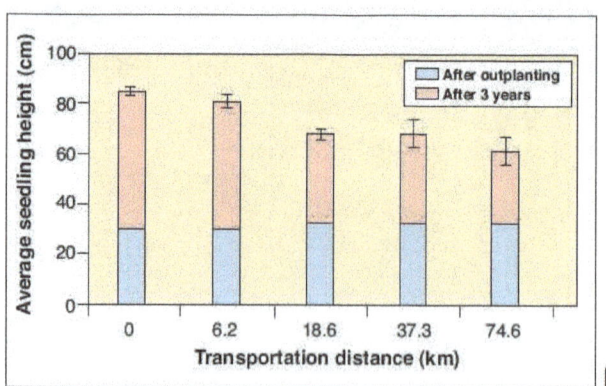

Figure 7.5.11—*Refrigerated vans ("reefers") are used to ship nursery stock for long distances (A). The risk of injury to nursery stock increases with the shipping distance (B). Dropoff reefer units can provide ideal long-term, onsite storage (C) (B, modified from Luoranen and others 2004).*

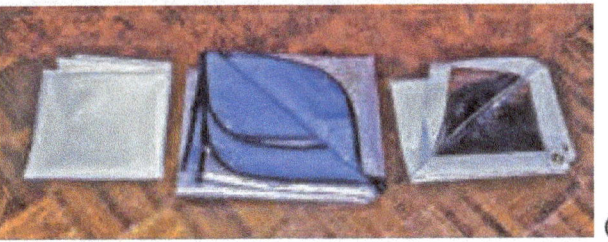
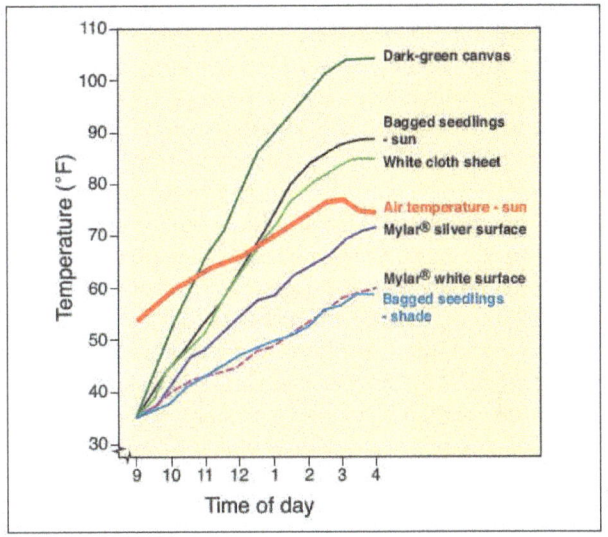

Figure 7.5.12—*Nonrefrigerated nursery delivery vehicles should be painted white and insulated to keep inside temperatures low (A). In open pickups, nursery stock should be covered with tarps (B); special reflective tarps (C) are commercially available. Research has shown that reflective Mylar® tarps provide much better insulation than standard green canvas ones (D) (D, modified from DeYoe and others 1986).*

refrigerated nursery stock ranged from 2 to 10 °C (36 to 50 °F) in refrigerated vehicles compared with 10 to 22 °C (50 to 72 °F) in nonrefrigerated vehicles (Stjernberg 1996). If boxes are loaded by hand, place spacers (such as wooden boards or foam blocks) between boxes or bags to allow air circulation and to prevent the load from shifting.

The Mt. Sopris Nursery in Colorado had specially designed reefer units developed that could be dropped off on the outplanting site to provide longer term, onsite storage (Figure 7.5.11C).

7.5.4.2 Delivery in nonrefrigerated trucks

For shorter trips, nonrefrigerated trucks are often used. Delivery vans should be aluminum or painted white to reflect sunlight and should be parked in the shade during stops and when they reach the outplanting site (fig. 7.5.12A). Plant ProTek is a new insulated truck liner that has been successfully tested with ornamental stock shipments and should have application for native plant nursery stock (Anonymous 2006). Adding "blue ice" in the boxes of small shipments would help keep temperatures down, although it could increase delivery costs.

Small pickup trucks. If open pickups must be used, then boxes of plants should be covered with a reflective tarp (fig. 7.5.12B). Specially constructed Mylar® tarps with white outer and silver inner surfaces are available from reforestation supply companies (fig. 7.5.12C). In operational trials, plants under such tarps were as cool as those stored in the shade (fig. 7.5.12D). Dark-colored tarps, such as army-green canvas, laid directly on boxes, however, allow plants to heat to damaging levels and should never be used (DeYoe and others 1986).

Commercial parcel trucks. Many State and private nurseries often ship small quantities of an assortment of native plants to a variety of customers. For example, the University of Idaho Frank Pitkin Nursery, which serves as the State nursery for Idaho, routinely ships an average of 120 seedlings at a time to each of its 1,500 customers throughout the state. To facilitate this, plants are extracted from containers and placed into plastic bags. Bags of plants are then bulked into stackable plastic tubs for cooler storage (fig. 7.5.13A). These tubs provide maximum flexibility in the cooler, allowing the configuration of the storage facility to be adjusted from year to year as production numbers vary. As the shipping season progresses, empty tubs can be removed from the cooler, thus creating additional workspace to process orders. As orders need to be shipped, employees move from tub to tub, gathering the appropriate species and quantities (fig. 7.5.13B). Completed orders are placed into cardboard boxes, weighed (Figure 7.5.13C), labeled (fig. 7.5.13D), and prepared for shipping by commercial delivery companies, such as United Parcel Service (UPS) or FedEx (fig. 7.5.13E). Because these packages will not always be handled by trained personnel, boxes must be heavy duty to protect nursery stock (fig. 7.5.13F). Within Idaho, all orders are usually delivered within 2 days. Customers are advised automatically by e-mail when their order leaves the nursery, and they receive tracking numbers to monitor the delivery process.

Figure 7.5.13—*Bundles of plants are taken from bulk bins in cooler storage (A), individual orders are assembled in cardboard shipping boxes (B), boxes are weighed (C), and shipping labels with bar codes are printed (D). Boxes are shipped to customers by package delivery early each week to ensure delivery before the weekend (E). The value of proper packaging becomes especially apparent with nursery stock delivered by commercial parcel services (F).*

7.5.5 Summary and Recommendations

After a crop begins the process of leaving the growing or storage area for the outplanting site, the financial and plant-quality risks peak—plants have reached their full economic value and should be at their highest quality level. Plants are living, perishable organisms and it is paramount to minimize stresses that can reduce their quality. The three primary types of stress that seedlings may encounter are moisture loss (desiccation), temperature extremes, and physical damage. Stock should be regularly monitored and handled gently to avoid exposure to stress. The effects of stress are cumulative—plants exposed to excessive stress may be dead at the time of outplanting or die shortly afterward. Unfortunately, the more common scenario is that the accumulation of stress causes a gradual and cumulative reduction in survival and growth that may or may not become apparent until weeks or months after outplanting.

The key to successful handling and shipping is to minimize stresses. Special equipment is often used to move plants that remain in their original containers throughout storage and shipping, thus reducing physical stress. Many nurseries, however, extract stock from containers and store and ship them in boxes or bags to reduce storage volume and shipping weight and to avoid the logistics of having containers returned to the nursery. In general, large numbers of plants, such as those shipped by reforestation, forest product, or Federal nurseries, are routinely shipped in large refrigerated trailers to reduce temperature stress and to maintain stock quality. For smaller nurseries, such as private native plant operations and State facilities, and for small quantities of stock that are being sent to locations near the nursery, stock is often shipped without refrigeration—such shipments are successful if care is taken to minimize extremes of temperature and physical stress.

7.5.6 Literature Cited

Adams, J.C.; Patterson, W.B. 2004. Comparison of planting bar and hoedad planted seedlings for survival and growth in a controlled environment. In: Connor, K.F., ed. Proceedings of the 12th biennial southern silvicultural research conference. Gen. Tech. Rep. SRS-71. Asheville, NC: USDA Forest Service, Southern Research Station: 423-424.

Anonymous. 2006. Greenhouse on wheels: new shipping technology converts dry vans into nursery stock haulers. Digger 50(1): 46-47.

Balneaves, J.M.; Menzies, M.I. 1988. Lifting and handling procedures at Edendale Nursery: effects on survival and growth of 1/0 Pinus radiata seedlings. New Zealand Journal of Forestry Science 18: 132-134.

Brown, K.E. 2007. Personal communication. Juniper, NB: J.D. Irving, Ltd., Juniper Tree Nursery.

DeYoe, D. 1986. Guidelines for handling seeds and seedlings to ensure vigorous stock. Special Publication 13. Corvallis, OR: Oregon State University, Forest Research Laboratory. 24 p.

DeYoe, D.; Holbo, H.R.; Waddell, K. 1986. Seedling protection from heat stress between lifting and planting. Western Journal of Applied Forestry 1(4): 124-126.

Fancher, G.A.; Mexal, J.G.; Fisher, J.T. 1986. Planting and handling conifer seedlings in New Mexico. CES Circular 526. Las Cruces, NM: New Mexico State University. 10 p.

Fox, T.R.; Jokela, E.J.; Allen, H.L. 2007. The development of pine plantation silviculture in the Southern United States. Journal of Forestry 105(7): 337-347.

Hanninen, H.; Pelkonen, P. 1989. Dormancy release in *Pinus sylvestris* L. and *Picea abies* (L.) Karst. seedlings: effects of intermittent warm periods during chilling. Trees 3(3): 179-184.

Kerr, G. 1994. A comparison of cell grown and bare-rooted oak and beech seedlings one season after outplanting. Forestry 67(4): 297-312.

Luoranen, J.; Rikala, R.; Smolander, H. 2004. Summer planting of hot-lifted silver birch container seedlings. In: Cicarese, L., ed., Nursery production and stand establishment of broadleaves to promote sustainable forest management, APAT, 2004. IUFRO S3.02.00, May 7-10, 2001, Rome, Italy: 207-218. http://www.iufro.org/publications/proceedings/ (accessed 23 January 2009).

McKay, H.M.; Gardiner, B.A.; Mason, W.L.; Nelson, D.G.; Hollingsworth, M.K. 1993. The gravitational forces generated by dropping plants and the response of Sitka spruce seedlings to dropping. Canadian Journal of Forest Research 23: 2443–2451.

Paterson, J.; DeYoe, D.; Millson, S.; Galloway, R. 2001. The handling and planting of seedlings. In: Wagner, R.G.; Colombo, S.J., eds. Regenerating the Canadian forest: principles and practice for Ontario. Markham, ON, Canada: Ontario Ministry of Natural Resources and Fitzhenry & Whiteside Ltd.: 325-341.

Rose, R.; Haase, D.L. 2006. Guide to reforestation in Oregon. Corvallis, OR: Oregon State University, College of Forestry. 48 p.

Self, A.B.; Ezell, A.W.; Guttery, M.R. 2006. First-year survival and growth of bottomland oak species following intensive establishment procedures. In: Connor, K.F., ed. Proceedings of the 13th biennial southern silvicultural research conference. Gen. Tech. Rep. SRS-92. Asheville, NC: USDA Forest Service, Southern Research Station: 209-211.

Sharpe, A.L.; Mason, W.L.; Howes, R.E.J. 1990. Early forest performance of roughly handled Sitka spruce and Douglas fir of different plant types. Scottish Forestry 44: 257–265.

Stjernberg, E.I. 1996. Seedling transportation: effect of mechanical shocks on seedling performance. Tech. Rep. TR-114. Pointe-Claire, QC, Canada: Forest Engineering Research Institute of Canada. 16 p.

Stjernberg, E.I. 1997. Mechanical shock during transportation: effects on seedling performance. New Forests 13(1-3): 401-420.

Tabbush, P.M. 1986. Rough handling, soil temperature, and root development in outplanted Sitka spruce and Douglas-fir. Canadian Journal of Forest Research 16: 1385–1388.

(USDA Forest Service) U.S. Department of Agriculture. 1989. A guide to the care and planting of southern pine seedlings. Southern Region, Mgt. Bull. R8-MB39. 44 p.

The Container Tree Nursery Manual

Volume Seven

Chapter 6
Outplanting

Contents

7.6.1 **Introduction** *154*
Limiting factors on the outplanting site
Timing of the outplanting window
Outplanting tools and techniques

7.6.2 **Outplanting Windows** *155*

7.6.3 **Onsite Handling and Storage** *157*
7.6.3.1 Inspecting nursery stock *157*
7.6.3.2 Hot-planted stock and open-stored stock *158*
7.6.3.3 Stock from refrigerated storage *158*
Thawing frozen stock
Outplanting frozen stock

7.6.4 **Preplanting Preparations** *162*
7.6.4.1 Check soil moisture and temperature *162*
7.6.4.2 Monitor air humidity, temperature, and wind speed *162*
7.6.4.3 Site aspect and planting sequence *162*
7.6.4.4 Watering plants and root dips *162*
7.6.4.5 Site preparation *163*
Scalping
Mounding
Inverting
Site preparation and frost heaving
7.6.4.6 Application of herbicides ("chemical scalping") *167*
7.6.4.7 Site preparation for restoration plantings *167*

7.6.5 **Selecting Plant Spacing and Pattern** *169*
7.6.5.1 Selecting planting spots *169*
Microsites

7.6.6 **Crew Training and Supervision** *171*
7.6.6.1 Plant handling *171*
7.6.6.2 Proper planting technique *171*

7.6.7 **Hand-Planting Equipment** *173*
7.6.7.1 Dibbles *173*
7.6.7.2 Bars *174*
7.6.7.3 Hoedads *175*
7.6.7.4 Shovels *176*
7.6.7.5 Tubes *176*
7.6.7.6 Motorized augers *176*

7.6.8 **Machine-Planting** *179*
7.6.8.1 Machines towed behind tractors *179*
7.6.8.2 Self-propelled planting machines *181*
Bräcke planting machine
M-Planter
Ecoplanter

7.6.9 **Planting Equipment for Large Stock** *184*
7.6.9.1 Expandable stinger *184*
7.6.9.2 Pot planter *185*

7.6.10 **Treatments at Time of Planting** *186*
7.6.10.1 Protection from animal damage *186*
7.6.10.2 Fertilization *187*
7.6.10.3 Mulches *188*
7.6.10.4 Shelters *188*
7.6.10.5 Shading *189*
7.6.10.6 Irrigation *189*

7.6.11 **Monitoring Outplanting Performance** *190*
Check the number and spatial distribution of plants
Aboveground inspection
Belowground inspection
7.6.11.1 What type of survey is best? *190*
Circular plots
Stake rows
7.6.11.2 What sampling design is best? *192*
7.6.11.3 How many plots are necessary? *192*

7.6.12 **Conclusions and Recommendations** *193*

7.6.13 **Literature Cited** *194*

7.6.1 Introduction

Outplanting is the final stage of the nursery process, but before we get to specific techniques, we should review some important concepts. Outplanting performance (survival and growth) depends on three factors, which are the final elements of the Target Plant Concept (fig. 7.6.1).

Limiting factors on the outplanting site. Each site is different, so it is critical to identify the environmental factors that can limit plant survival and growth. Temperature and moisture are usually the most limiting and are discussed in Section 7.6.4. Other site factors, such as aspect and soil type, must also be considered. Sites with south or southwest aspects will dry out more quickly and should therefore be planted first. In some cases, shade materials may be required. Some planting tools should not be used on fine-textured soils, such as silts and clays; this will be discussed in Section 7.6.7.

Outplanting sites must be evaluated well in advance of the actual outplanting. Although the site evaluation process will not be covered in detail here, two good resources exist. First, the Forest Service requires a detailed Reforestation Plan for each planting project (USDA Forest Service 2002). Second, a very comprehensive example of how to conduct a site evaluation on a restoration site can be found in Steinfeld and others (2007). Because of the highly disturbed nature of restoration sites, site evaluation is even more critical before planting can begin (Munshower 1994).

Timing of the outplanting window. For each site, there is an ideal time to plant, and the process for determining this "window" is discussed in Section 7.6.2.

Outplanting tools and techniques. The processes for selecting the best way to plant nursery stock are discussed in Sections 7.6.3 to 7.6.9, and Section 7.6.10 describes how to evaluate the quality of the outplanting project.

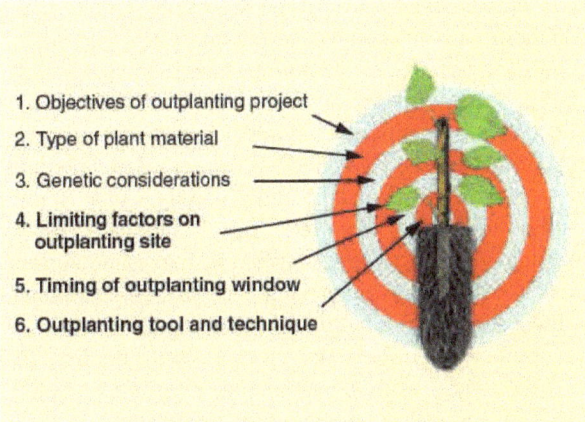

Figure 7.6.1—*The final three steps of the Target Plant Concept are critical to outplanting success and should be considered when planning and initiating outplanting projects.*

7.6.2 Outplanting Windows

Years of experience have proved that the best time to outplant seedlings is when they are dormant and least susceptible to the stresses of harvest, storage, shipping, and planting. The outplanting window concept was introduced as a critical part of the Target Plant Concept (see Section 7.1.1.5) and is defined as the period of time during which environmental conditions on the site most favor survival and growth of nursery stock. Traditionally, outplanting windows were established by harvesting nursery stock and observing outplanting performance. Plant quality tests, such as cold hardiness, are also good ways to determine when nursery stock is most hardy and best able to survive the stresses of outplanting. For example, cold hardiness testing of ponderosa pine (*Pinus ponderosa*) and Douglas-fir (*Pseudotsuga menziesii*) over a 4-year period shows how the duration of outplanting windows varies from year to year (fig. 7.6.2A).

The start and end dates of an outplanting window are constrained by limiting factors on the planting site. Soil moisture and temperature are the usual constraints on most sites, but other environmental or biological factors can also limit plant survival and growth (see Section 7.1.1.4). For high-value plantings where irrigation can be supplied, container stock can be outplanted throughout the year in appropriate weather conditions and with proper handling (White 1990). Changing weather patterns have caused changes in the outplanting window. In eastern Texas, an extended drought has caused foresters to change from traditional spring outplanting with bareroot stock to fall outplanting with container stock. Tests show fall-planted container seedlings had a 93 percent survival rate compared with 67 percent for bareroot stock (Taylor 2005).

In most of the continental United States, nursery stock is outplanted during late winter or early spring, when soil moisture is high and evapotranspirational losses are low. In most of Canada and the United States, this typically occurs from January to April for lower elevations (fig. 7.6.2B). These outplantings have used dormant stock that was harvested during early winter and stored for 2 to 8 months under refrigeration or in outdoor compounds (see Chapter 4 in this volume for more information).

At high elevations and latitudes, however, it is impossible to plant during late winter or early spring, because persistent snow cover keeps soil temperatures low and limits access. This means that all nursery stock must be outplanted during a relatively short window when long days and high solar angles cause high evapotranspiration rates (fig. 7.6.2C). Therefore, some foresters in northern Canada, Scandinavia, and the northern mountains of the Western United States have outplanted container stock during early summer or even later in the fall (Luoranen and others 2004; Page-Dumroese and others 2008; Tan and others 2008). Container plants have a wider outplanting window because they suffer less transplant shock; their roots are protected in the plug and not damaged during harvesting. In addition, with modern container nursery techniques, it is possible to culture plants to better tolerate outplanting stresses. Because nursery stock outplanted during summer or fall is not dormant, this is known as "hot-planting." Hot-planted stock needs some hardening to withstand the stresses of harvesting, shipping, and outplanting; moisture stress or short-day ("blackout") treatments are most commonly used (Landis and Jacobs 2008). Finnish researchers have been conducting outplanting research on hot-planted Norway spruce (*Picea abies*) and silver birch (*Betula pendula*) for several years (Louranen and others 2005). For example, to investigate the effect of drought on outplanting performance, hot-lifted Norway spruce seedlings were subjected to up to 6 weeks of water stress in a research plot (Helenius and others 2002). They found that hot-lifted stock with wet plugs that were outplanted in July had better root egress than those planted later that year, or even those that were stored and outplanted the following spring (fig. 7.6.2D).

Therefore, many biological and operational factors must be considered when determining the best outplanting window, but nothing substitutes for actual experience, and survival and growth are always the best guides. (For more information on hot-planting see Sections 7.4.2 and 7.6.3.2.)

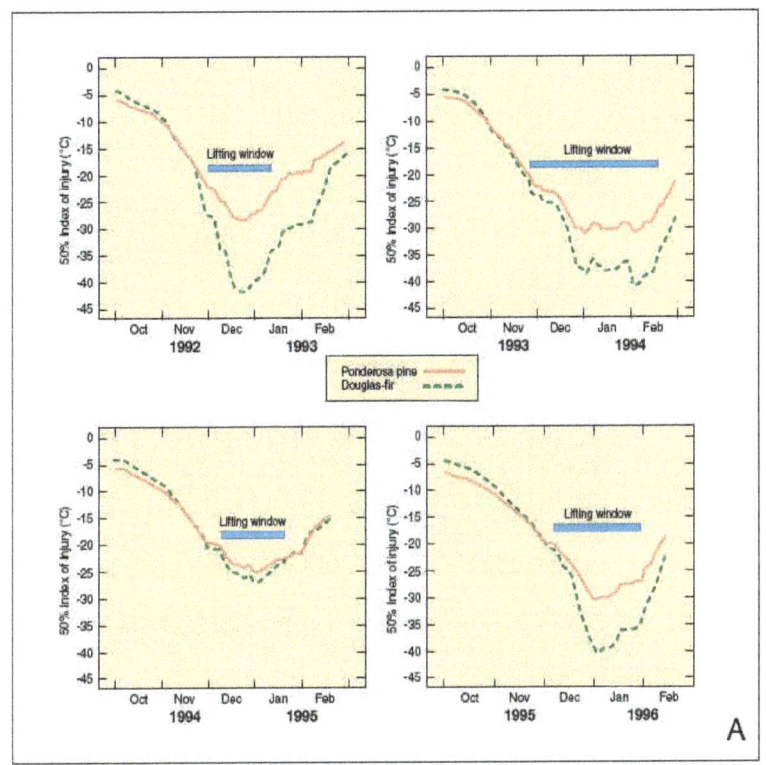
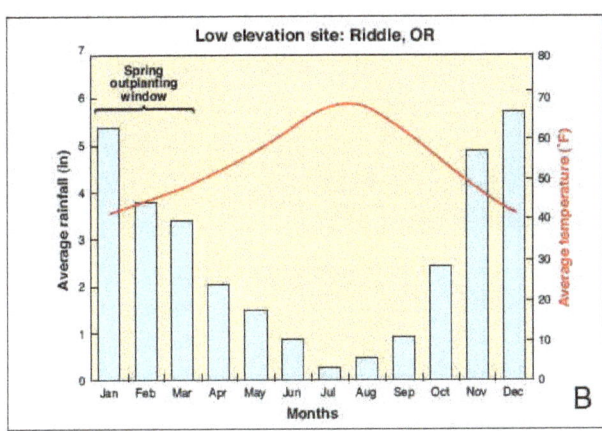
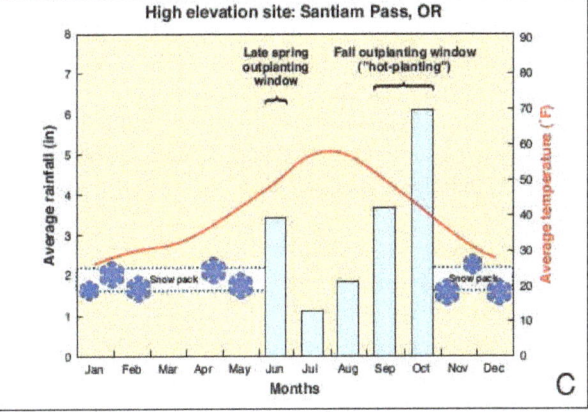
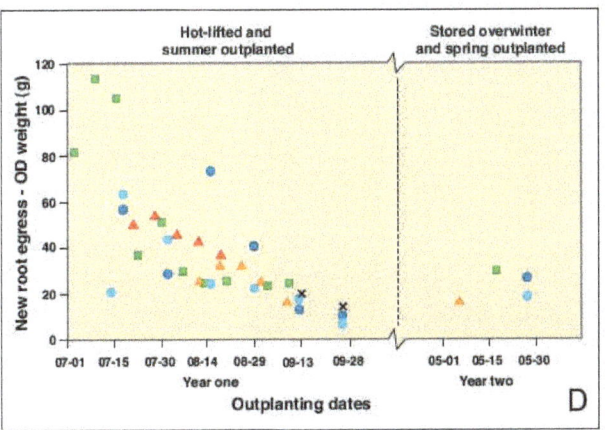

Figure 7.6.2—Outplanting windows are established from observations of lifting and planting successes or with plant-quality testing (A). In most of the United States and Canada, the outplanting window occurs during late winter or early spring (B). At higher elevations and latitudes, however, the outplanting window is later spring or early summer due to persistent snow and cold soil temperatures (C). Hot-lifted Norway spruce outplanted in early summer had more root egress than those planted later that year, or even overwintered stock planted during the traditional spring outplanting window (D) (A, modified from Tinus 1996; B&C, courtesy of Steinfeld and others 2007; D, modified from Louranen and others 2006).

7.6.3 Onsite Handling and Storage

Nursery stock should be outplanted as soon as it arrives on the project site, but that is often operationally impossible. Weather delays, worker scheduling, and poor communication are just a few of the reasons why onsite storage may be necessary. The duration of onsite storage should last for only a few days, although, under unanticipated weather, such as heavy snow, it can reach a week or more. Therefore, it is always wise to plan ahead. Ideally, project managers should bring only as much stock as can be planted on a given day to avoid the need for onsite storage. Distance and other logistical factors, however, may make this difficult.

Overheating and desiccation are the major stresses that can occur during onsite storage. Because of significant differences in dormancy stage and hardiness, however, nursery stock for hot-planting must be treated differently from stock that comes from refrigerated storage.

7.6.3.1 Inspecting nursery stock

As discussed in Chapter 7.5, many things can happen between the harvest and outplanting of nursery stock. Therefore, it is a good idea to conduct a thorough inspection of nursery stock when it arrives at the outplanting site. All boxes should be opened and checked for the following (Mitchell and others 1990):

- In-box temperatures of refrigerated stock should be checked upon delivery (fig. 7.6.3A) and should be cool, no warmer than 2 to 4 °C (36 to 39 °F). Stock delivered in containers or hot-plant stock should be kept as cool as possible and out of direct sunlight.
- If possible, use a pressure chamber to check plant moisture stress of a sample of plants (fig. 7.6.3B). (Target PMS values can be found in Chapter 7.2.)
- Nursery stock should not smell sour or sweet, which is evidence that the stock has been too warm or excessively wet.
- Root plugs should be moist. If the plants have foliage, most often it should be a healthy green. For species with terminal buds, those buds should still be firm.
- Check the firmness of the bark around the root collar. The bark should not easily slough off and the tissue underneath should be creamy, not brown or black, which indicates frost injury.

Figure 7.6.3— *Nursery stock should be inspected upon delivery to the outplanting site. Check in-box temperatures of boxed plants (A) and, if possible, measure the plant moisture stress with a pressure chamber (B). Storage molds can become a serious problem in onsite storage, so check for gray or colored mycelia within the foliage (C).*

- Spread the foliage to check for white or gray mycelia (fig. 7.6.3C), which is evidence of storage molds, such as *Botrytis cinerea*. In particular, check foliage at the base of the crown. If mold is present, check the firmness of the tissue underneath. Soggy or water-soaked tissue indicates serious decay and those plants should be culled. Plants with superficial mycelia without corresponding decay should be planted immediately. Fungal molds will not survive after exposure to ambient conditions on the site.

7.6.3.2 Hot-planted stock and open-stored stock

Hot-planted stock, because it is not fully dormant or hardy, should be outplanted immediately or stored on the project site for only a day or two. The key to a successful hot-planting operation is careful planning and coordination between the nursery and planting project managers. Ideally, hot-plant stock should be packed upright in cardboard boxes without plastic bag liners that can reduce air exchange and increase respirational heat buildup. If stock is pulled and wrapped, using white packing boxes will help to reflect sunlight and keep in-box temperatures lower (Kiiskila 1999).

At the outplanting site, tops of cardboard boxes containing open-stored or hot-planting stock should be opened to dissipate heat and promote good air exchange. If not already so, the plants should be set upright and placed in a shady area as soon as they arrive on site. Unfortunately, trees and other natural shade are absent on many reforestation and restoration sites, but even when natural shade is available, it can be difficult to keep plants in the shade all day (fig. 7.6.4A). Therefore, plan on erecting some type of artificial shade. Tarps or shadecloth suspended between poles is effective (fig. 7.6.4B). As shown in figure 7.5.12D in the previous chapter, dark-colored tarps absorb and reradiate solar heat (Emmingham and others 2002); therefore, canvas tarps should be suspended above the nursery stock to allow for good air circulation. Wetting the tarps regularly will keep the air cooler through evaporative cooling (Mitchell and others 1990).

Moisture stress is another concern with open-stored or hot-planted stock because plants are transpiring during delivery and onsite storage. As with respiration, the transpiration rate is a function of temperature, but sunlight intensity is equally important. Therefore, it makes sense to check that root plugs are fully charged and plants are not under any moisture stress immediately before outplanting. Irrigating container plants on the project site is not commonly done but recent research with hot-planted birch (*Betula* spp.) and spruce (*Picea* spp.) seedlings showed that water before outplanting significantly increased survival (fig. 7.6.4C). So, the best onsite storage for open-stored or hot-planted nursery stock has access to a reliable water source (fig. 7.6.4D) because frequent watering requires large volumes of water (Mitchell and others 1990).

7.6.3.3 Stock from refrigerated storage

Nursery plants delivered from cooler or freezer storage must be treated differently than open-stored or hot-planted stock because they are still fully dormant and hardy and, ideally, should be kept that way until outplanting. So, whenever possible, refrigerated trucks ("reefers") should be used for transportation to the site as well as for onsite storage (fig. 7.6.5A). Each truck should receive a mechanical check before use, and the storage van should be precooled by operating the compressor for at least 4 to 6 hours (Paterson and others 2001). Anticipate mechanical failure by having a backup plan.

Snow caches, culvert or pole structures covered with snow and an insulating material such as sawdust or straw (fig. 7.6.5B), have been successfully used for onsite storage where conditions permit (Paterson and others 2001). In a Canadian trial, a custom-made, insulated storage building was effective in protecting container stock from both frost damage and overheating (Zalasky 1983).

Boxes or bags of stock, stored either under refrigeration or in insulated buildings, should be patched if torn during shipping and handling and kept closed. The temperature inside the boxes or bags can be much warmer than the outside temperature because plants produce heat during respiration. As the temperature increases, so does the rate of respiration, which further increases the temperature. Therefore, the temperature in boxes or bags should be monitored on delivery and daily thereafter (fig. 7.6.3A). Make sure that the in-box temperature remains above freezing but below 10 °C (50°F) (Rose and Haase 2006). If nursery stock is exposed to warm temperatures for an

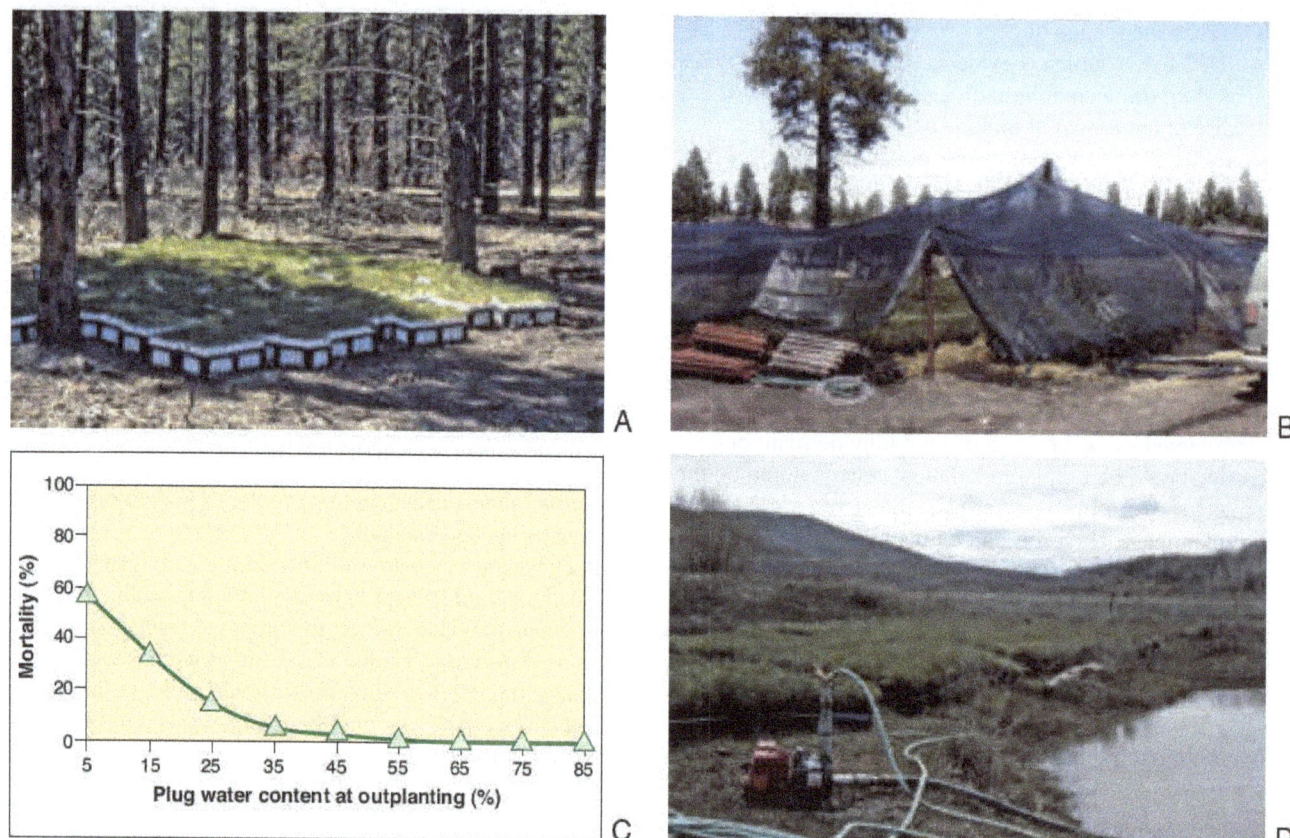

Figure 7.6.4—*All nursery stock should be kept in the shade on the outplanting site, but natural shade moves with the sun (A). Artificial shade from tarps or shadecloth is needed on many project sites (B). Watering plants immediately before outplanting has proven beneficial for hot-planted silver birch on dry sites (C), so provide for a source of irrigation water (D) (C, modified from Luoranen and others 2004).*

extended period, standard seedling quality tests (root growth potential, chlorophyll fluorescence, and cold hardiness) and in-container concentration of ethanol were shown to accurately predict seedling performance (Maki and Colombo 2001). (See Chapter 7.2 for more information on plant quality tests.)

It is also prudent to check inside a few boxes for signs of storage molds such as *Botrytis cinerea* (fig. 7.6.3C). This common nursery pest can increase rapidly after refrigerated storage, perhaps because of increased carbon dioxide levels inside boxes and bags (see Volume Five for more information).

Thawing frozen stock. Plants with root plugs frozen together must be thawed before outplanting. Some customers want their stock thawed before shipping by either "rapid" or "slow" thawing (fig. 7.6.6A). However, the definitions of "slow" and "rapid" vary considerably (table 7.6.1). Originally, slow thawing was considered best (for example, Mitchell and others 1990) and was typically done at the nursery. Recent research comparing the two thawing techniques found no differences after two growing seasons (Rose and Haase 1997). In the most comprehensive physiological research on thawing frozen stock (Camm and others 1995), cold hardiness tests showed that rapidly thawed stock was more hardy and also resumed shoot growth earlier than

Figure 7.6.5—Nursery stock from refrigerated storage should be kept in reefer trucks (A), insulated structures, or snow caches (B) until outplanting.

Figure 7.6.6—Frozen nursery stock must be carefully thawed at warm temperatures for 24 to 48 hours (A). Never expose frozen plants to direct sunlight (B), but open boxes or bags (C) in a shady location (D) (A, modified from Paterson and others 2001.)

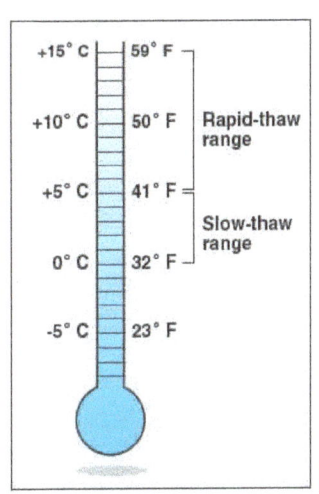

Table 7.6.1—*Common thawing regimes for frozen container nursery stock*

Speed of thawing	Reference	Temperatures	Duration
Slow thaw	Camm and others (1995)	5 °C (41 °F) followed by 15 °C (59 °F)	7 days 2 days
	Rose and Haase (1997)	0 to 3 °C * (32 to 37 °F)	42 days
	Kooistra and Bakker (2002)	0 to 3 °C * (32 to 37 °F)	21 to 35 days
Rapid thaw	Camm and others (1995)	22 °C (72 °F)	2 to 5 hours
	Rose and Haase (1997)	7 °C (45 °F)	5 days
	Kooistra and Bakker (2002)	5 to 10 °C (41 to 50 °F)	5 to 10 days

* Operational cooler storage conditions

slowly thawed seedlings. Moreover, shoot and root growth measurements after 3 months were similar. These results suggest that a good operational procedure might be to remove bundles of frozen stock from shipping containers and lay them on the ground or to open shipping boxes or bags (fig. 7.6.6C) in a well-ventilated shady location. Never attempt to thaw frozen nursery plants by placing them in direct sunlight (fig. 7.6.6B), as this can cause serious moisture and temperature stress. Do not physically pry frozen root plugs apart because this can cause serious damage (Mitchell and others 1990). Defrost only enough stock that can be planted in a couple of days. The ideal situation is to setup a thawing operation in which frozen stock is removed from refrigerated storage and then thawed in an adjacent shade structure (fig. 7.6.6D).

Outplanting frozen stock. Outplanting nursery stock with frozen root plugs would save the time and effort needed to thaw plants. A few years ago this was not possible because root plugs were frozen together; now technology for packing singulated plants is available. Results of field trials, however, are mixed. In British Columbia, the performance of western larch (*Larix occidentalis*), lodgepole pine (*Pinus contorta*), and interior spruce planted frozen was not significantly different from thawed plants 2 years after outplanting (Kooistra and Bakker 2005). Other studies indicate that site conditions have an overriding effect. In an outplanting study of Norway spruce seedlings in Finland, thawed seedlings outperformed frozen stock in survival and shoot and root growth in warm and cold soils (Helenius 2005). In a more recent trial, the physiological processes of thawed and frozen Douglas-fir container seedlings that were exposed to either "cool and moist" or "warm and dry" conditions were monitored. Thawed plants had higher photosynthesis rates and more active buds and roots than plants that were planted frozen, which could affect subsequent outplanting performance (Islam and others 2008). Obviously, more research trials under a wide variety of outplanting site conditions are needed before outplanting frozen stock can be recommended.

7.6.4 Preplanting Preparations

Before the outplanting actually begins, several preparations should be made to ensure the project runs smoothly and successfully.

7.6.4.1 Check soil moisture and temperature

Soil moisture plays a vital role in the uptake and translocation of nutrients and can have a significant influence on plant survival and growth (Helenius and others 2002). Following outplanting, a root system must be able to take up sufficient water from the surrounding soil to meet the transpirational demands of the shoot. If soil moisture is inadequate, the newly planted seedling can become stressed, resulting in reduced growth and increased mortality. Lower photosynthetic rates can occur in newly planted seedlings under water stress, which results in lower root regenerating ability (Grossnickle 1993). Ideally, soil water potential in the top 25 cm (10 in) should be greater than –0.1 MPa at the time of outplanting (Cleary and others 1978; Krumlik 1984).

Soil temperature has a profound effect on root development (Balisky and Burton 1997; Domisch and others 2001; Landhäusser and others 2001). The ideal soil temperature range for root growth is 5 to 20 °C (41 to 68 °F) (fig. 7.1.5B), so planting may have to be delayed until soil temperatures increase. When transpirational demands are high but cold soils limit water uptake, plants may experience a "physiological drought" that can limit survival and growth (Mitchell and others 1990). In Ontario, planting projects are not started until soil temperatures exceed 5 °C (41 °F).

7.6.4.2 Monitor air humidity, air temperature, and wind speed

Weather conditions at the time of outplanting have a direct effect on plant moisture stress. Although an increase in both air temperature and wind speed affect transpiration, wind effects are more difficult to quantify. Conditions become critical when air temperatures exceed 25 °C (78 °F) and relative humidity is lower than 30 percent (Paterson and others 2001). Relative humidity does not influence evapotranspiration rates as much as vapor pressure deficit, which is the difference between the amount of water the air can hold at a given temperature and the amount of water at saturation. Sample calculations can be found in Cleary and others (1978).

Therefore, planting is best done during the early morning hours when air temperatures are cool and wind speeds are low. When weather is sunny, windy, or dry it is necessary to take extra protective precautions to minimize plant stresses. In extreme cases, the planting operation may have to be suspended.

7.6.4.3 Site aspect and planting sequence

Conditions will vary at different locations in the planting area, especially in mountainous terrain. Aspect, or direction of solar exposure of mountain slopes, is one of the most important factors affecting outplanting success. South- and west-facing aspects have a hotter, drier environment than north and east aspects and should be planted first. Shading of outplanted stock is often required on these aspects (see following section). Deer and elk often use southern and western exposures for winter range so this impact must also be considered (USDA Forest Service 2002).

Be sure to consider access and transportation distance from the on-site storage. It is generally a good idea to start at the furthest location and plant back towards access roads.

7.6.4.4 Watering plants and root dips

The practice of dipping plant roots to protect them from stress during outplanting has been around for many years because it is intuitively attractive, especially for bareroot stock. Roots of nursery plants dry as they are exposed to the atmosphere during harvesting and handling, so it makes sense to rehydrate them or apply a coating to protect them (Chavasse 1981). Many different commercial root dips have become available and most are superabsorbent hydrogels. These crosslinked polymers can absorb and retain many times their own weight in water and are routinely applied to bareroot stock as root dips. Little research on the benefits of hydrogels to container stock has been published. However, one recent trial with Eucalyptus seedlings with root plugs dipped in a hydrogel slurry had significantly lower mortality at 5 months after outplanting compared with the controls. The author attributed this to increased soil moisture or contact between the root plug and the field soil (Thomas 2008). It would be interesting to see more research on this subject.

The plug should already be moist when it is unpacked. If not, then a quick water dip should be adequate to protect the roots from desiccation, because, as demonstrated in figure 7.6.4C, irrigating root plugs before outplanting has proved beneficial.

7.6.4.5 Site preparation

Trees and other native plants vary in their requirements for sunlight and other site resources to successfully regenerate. Site preparation (referred to as "site prep") to remove competing vegetation and site debris has several benefits (USDA Forest Service 2002). Biologically, it improves nursery stock survival and growth by reducing the competition from existing plants for nutrients, water, and sunlight. Roots from existing plants may have already occupied the soil profile and can easily reduce survival of outplanted nursery stock (fig. 7.6.7A). Operationally, site prep makes the physical process of planting easier by reducing surface debris on the site and removing the duff or sod layer. Removal of woody and herbaceous plants around Douglas-fir (*Pseudotsuga menziesii*) seedlings resulted in up to three times the stem volume after 8 years as compared with seedlings without vegetation control (Rose and Rosner 2005).

Site prep done at the time of outplanting can be accomplished by mechanical (scalping or mounding) or chemical (see Section 7.6.4.6) means.

Scalping. Scalping is the physical removal of grasses, forbs, small shrubs, and organic debris around planting holes (fig. 7.6.7B)—it is ineffective against larger woody plants that are too difficult to remove. Removing organic debris around the planting hole ensures that roots are in contact with mineral soil. Nursery stock planted in organic matter or duff dry out rapidly and often die (Grossnickle 2000). Scalping can also reduce the frequency of drought damage because of the reduction in competition (Barnard and others 1995; Nilsson and Orlander 1995). When light is the limiting factor, however, scalping can reduce growth because of reduced moisture and available nutrients (Miller and Brewer 1984).

Scalping can be accomplished with some planting tools such as the side of a hoedad blade (fig. 7.6.7C). With other planting implements such as augers, scalping is done beforehand by another worker. Planting contracts often contain specifications for the size and depth of scalping. For example, the Forest Service requires that all vegetation be removed from an area 30 to 60 cm (12 to 24 in) around the planting hole and 2 to 5 cm (1 to 2 in) in depth. On dry exposed sites, duff, litter, and rotten wood should be placed back on the cleared surface to serve as mulch (USDA Forest Service 2002). Scalping can definitely slow down planting productivity, but operational experience in Oregon found that a good hoedad planter can still scalp and plant 850 trees/day in Oregon (Henneman 2007).

Continuous scalping ("discing" or "scarifying") is done with tractor-drawn or self-powered equipment. The Bräcke Scarifier is mounted on the front of a prime mover on a three-point hitch allowing the operator to select individual spots. Two side-by-side scalps are about 2.5 m (8 ft) apart with about 2 m (6.5 ft) spacing between rows. Depending on terrain and desired density of the scalps, production varies from 0.5 to 2.0 ha (1.2 to 4.8 acres) per hour (Converse 1999). Discing, which produces a shallow furrow about 0.6 to 0.9 m (2 to 3 ft) and 5 to 10 cm (2 to 4 in) deep, has proved essential for establishment of longleaf pine (*Pinus palustris*) on abandoned farmland in the Southeastern United States (Shoulders 1958). Barnard and others (1995) found continuous scalping to be beneficial for these reasons:

- Reduced weed competition.
- Improved soil moisture availability.
- Less damage by root pathogens and insects.
- Increased planting efficiency.

The beneficial effects of scalping will vary on a site-by-site basis, and whether or not to scalp should be determined during the planning phase of any outplanting project. On grass-dominated sites in interior British Columbia, scalping was found to reduce evapotranspiration and increase soil moisture, which yielded better survival and growth of conifer seedlings (fig. 7.6.7D). In Oregon, increasing scalp size resulted in significant improvement in stem volume after 4 years (Rose and Rosner 2005). On the other hand, on boreal reforestation sites in northern British Columbia where plant competition is not severe, the additional time and expense of scalping failed to improve outplanting performance (Campbell and others 2006).

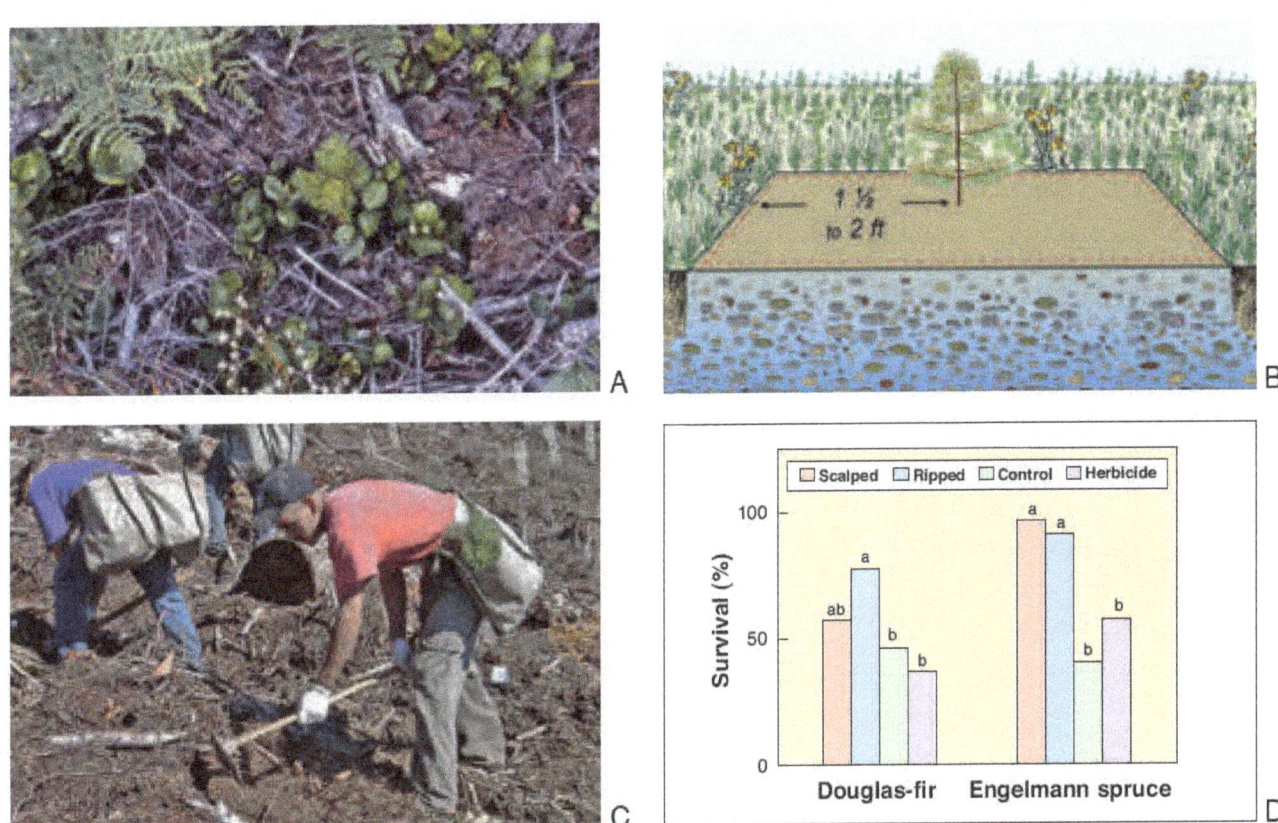

Figure 7.6.7—*Existing plants compete with outplanted nursery stock for moisture (A); scalping is the physical removal of plants and organic debris from around the planting hole (B). Spot scalping can be done with a planting tool, such as this hoedad (C). On a grass-dominated reforestation site, scalping improved conifer seedling survival compared with ripping, herbicide, or the no-treatment control (D) (D, modified from Fleming and others 1998).*

Mounding. On boreal and other cool-temperate planting sites, slow organic matter decomposition creates a heavy surface layer of duff that can be an impediment to planting nursery stock. "Mounding" is a general term for a type of site preparation that can treat several potentially limiting factors: plant competition, cold soil temperatures, poor aeration on wet sites, and nutrient deficiencies. The term mounding has been, however, applied to a variety of mechanical site treatments that can have widely different biological consequences. Sutton (1993) provides a thorough discussion of mounding and how it has been used worldwide.

For our purposes, we define mounding as the mechanical excavating and inverting of soil and sod to create separate piles that are higher than the existing terrain. With thick duff layers, the resulting mounds consists of a mineral soil cap over a double layer of humus (fig. 7.6.8A). While mounding was originally done by hand; a number of mechanical implements have been developed to speed up the process. For example, the Bräcke Mounder is a scarifier featuring a hydraulically operated entrenching spade followed by a mounding tool that uses soil from the scalped area. Widely used in Canada and Scandinavia, this machine produces mounds 16 to 26 cm (6 to 10 in) high with 3 to 19 cm (1 to 7 in) caps of mineral soil. Other studies have used modified mouldboard plows to generate a continuous mounded ridge (Sutton 1993).

The results of mounding have been generally favorable, at least in the short term. For example, compared with scarification and herbicide treatments, mounding produced

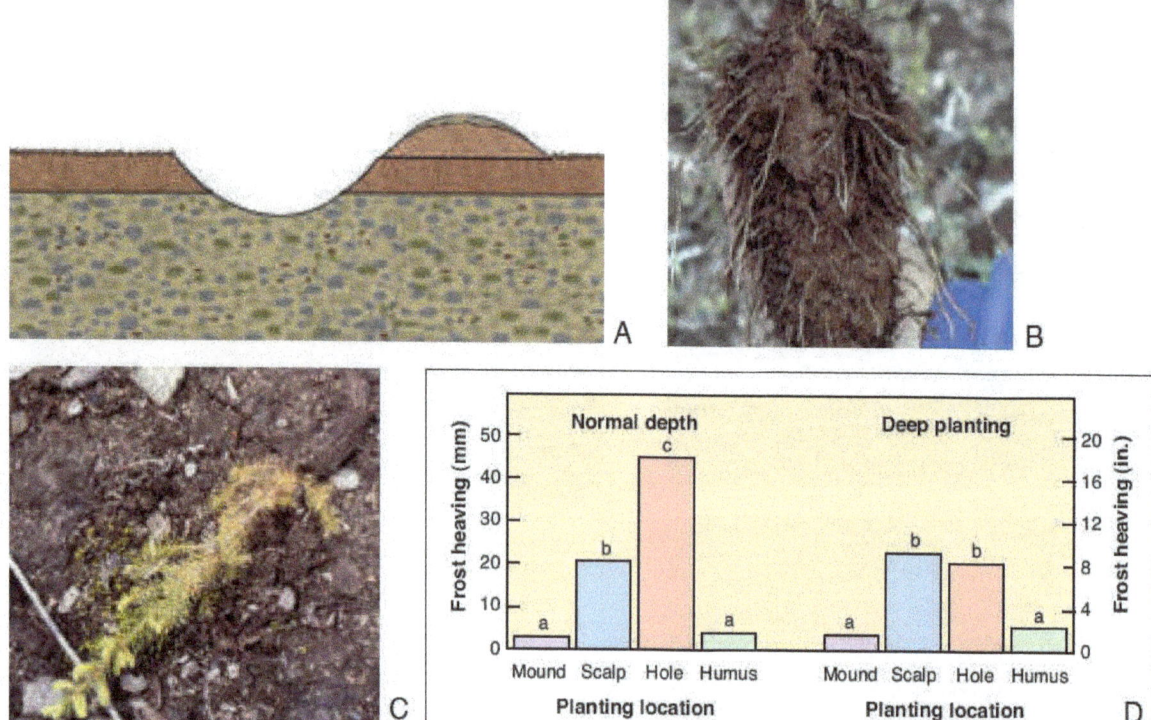

Figure 7.6.8—*On boreal sites with heavy duff layers or in waterlogged soils, mounding has proven to benefit plant survival and growth (A). Plants that can rapidly grow new roots (B) will be less susceptible to frost heaving (C). Mounding has also proven effective when the plants are positioned on top of the mound and not in the hole (D). Inverting achieves some of the same benefits as mounding but has a much more acceptable appearance (E) (B, courtesy of Cheryl Talbert; D, modified from Sahlen and Goulet 2002).*

strong, consistent, positive results for jack pine (*Pinus banksiana*) on grass-dominated sites (Sutton and Weldon 1993). Most research involved conifers, but a recent study found that mounding was an effective alternative to herbicides for establishing pedunculate oaks (*Quercus robur*) on waterlogged sites (Lof and others 2006). Conversely, Sutherland and Foreman (2000) found that mound planting resulted in less growth of black spruce (*Picea mariana*) compared with repeated herbicide treatments. Mounding has also been shown to help reduce injury by the European pine weevil (*Hylobius abietis*), which is the major regeneration pest in northern European forests. Because it reduces feeding damage by the weevil, mounding is common on 20 percent of Norway spruce plantations in Finland (Heiskanen and Viiri 2005).

Mounding has been criticized from an aesthetic and ecological standpoint and can have a negative effect on other forest values, such as recreation (Lof and others 2006). So, as with all site preparation treatments, mounding needs to be carefully evaluated on a site-by-site basis and compared with other site preparation options.

Inverting. This relatively new mechanical site preparation method uses an excavator to create planting spots containing inverted humus covered by loose mineral soil

Figure 7.6.9—*When competing vegetation is killed with herbicides prior to planting (A), the soil moisture that would have been lost to transpiration is conserved on the site (B).*

without making large mounds or ridges (fig. 7.6.8E). Research in Sweden with Norway spruce and lodgepole pine found that inverting produced significantly greater survival and stem volume growth after 10 years compared with plowing, mounding, disc trenching, or no scarification (Orlander and others 1998). A subsequent research trial with Norway spruce confirmed that inverting produced increased seedling survival compared with mounding or unscarified controls. Appearance and environmental effects were also measured and, compared with mounding, inverting reduced the alteration of the ground contour from 40 to 15 percent (Hallsby and Orlander 2004).

Site preparation and frost heaving. Frost heaving of recently outplanted nursery stock is a major problem on sites subject to repeated cycles of freezing and thawing. Heaving is a purely mechanical process whereby plants or other objects are slowly racheted out of the soil by repeated freezing and thawing (Goulet 1995). All nursery plants can be frost heaved, but, because of their smooth-walled root plugs, container plants are particularly susceptible.

Sites prone to frost heaving have high soil moisture and soil textures with good hydraulic conductivity (Bergsten and others 2001). The tendency to frost heave increases as pore size decreases, so silt and clay soils are most problematic. Southerly or southwesterly sites have more of a problem with frost heaving because the high solar exposure intensifies the freeze-thaw cycle.

The physiological condition of stock at outplanting can have a significant effect on frost heaving. Plants that have rapid root egress (fig. 7.6.8B) will become physically anchored into the soil and therefore less susceptible. Fear of frost heaving is a major reason why late fall outplanting is discouraged. Nursery stock that is outplanted so late that new roots cannot anchor the plant will be vulnerable to frost heaving (fig. 7.6.8C). In one study, however, stock outplanted later did not suffer more damage than stock outplanted earlier (Sahlen and Goulet 2002).

Site preparation treatments have a significant effect on frost heaving. Scarifying the planting spot increases potential for heaving, because the insulating humus layer and surrounding vegetation are removed, allowing diurnal temperatures to fluctuate more widely. On the other hand, mounding should reduce frost heaving because it provides better drainage and reduces capillary water rise (Bergsten and others 2001). Research on the effect of planting position on frost heaving showed that heaving was considerably higher in the hole where water migrated to the surface and froze into layers that attached to the plant. On top of mounds, frost heaving was as low as in the nontreated humus layer (fig. 7.6.8D). Although deep planting has been suggested as a way to provide better anchorage, it was ineffective in this study (Sahlen and Goulet 2002).

7.6.4.6 Application of herbicides ("chemical scalping")

Mechanical scalping is a time-consuming and therefore expensive site-preparation treatment. Another option is to kill competing vegetation around planting spots with herbicides in advance of the actual outplanting. A general-purpose herbicide like glyphosate (Roundup®) kills all types of plants in the treated area but has very low environmental effects and no residual activity. By killing competing plants before the planting project begins, soil moisture that would otherwise be lost by their transpiration (fig. 7.6.9A) is conserved onsite and will be readily available to the outplanted nursery stock (fig. 7.6.9B). On reforestation sites in Northern California mountains, hexazinone (Velpar®) herbicide is applied 1 to 2 years before the planting project to kill brush and other competing vegetation (Fredrickson 2003). Two years of intensive vegetative control was essential to successful reforestation on Weyerhaeuser lands in Washington State (Talbert 2008).

Concerns about phytotoxicity of sulfometuron methyl (Oust XP®) were addressed for Douglas-fir, western hemlock (*Tsuga heterophylla*), and western redcedar (*Thuja plicata*) container seedlings in coastal Oregon. Although root egress was initially restricted due to the herbicide, no significant effects were observed after 9 to 21 months, showing that any phytotoxicity was short lived (Burney and Jacobs 2009).

Herbicides not only eliminate transpirational water loss, but create a mulch of dead organic matter that reduces surface evaporation. Vegetation control with herbicides has been shown to increase subsequent survival and growth of outplanted nursery stock. An experiment evaluating three levels of vegetation control with chemical scalping significantly increased stem volume, basal diameter, and height of seedlings on four of five sites with increasing area of weed control, and the magnitude of difference between treatments increased with time (Rose and Ketchum 2002). Herbicide applications can also be effective in reducing fire hazards and eradicating non-native plants.

The best herbicide application method depends on the type of project. Aerial application with helicopters is efficient and cost effective for large reforestation or restoration projects. For forest plantation projects, herbicides can be sprayed in rows by all-terrain vehicles (ATVs) or by sprayers attached to ripping equipment. For smaller projects, herbicides can be applied with a backpack sprayer by a person trained in the selection of likely planting spots.

7.6.4.7 Site preparation for restoration plantings

On restoration planting sites, severe disturbance may require unusual site preparation to create suitable planting spots. After the eruption of Mount St. Helens in Washington State, the restoration of 60,700 ha (150,000 acres) of timberland posed some serious challenges (fig. 7.6.10A). Experiments showed that seedlings must be planted in mineral soil to survive, which required digging through 30 to 60 cm (1 to 2 ft) of volcanic ash at each planting spot (fig. 7.6.10B). In many cases, planting sites must undergo major stabilization before planting can occur. Because of their steep slopes and the erosive power of water, stream banks must be stabilized with bioengineering structures before they can be revegetated (fig. 7.6.10C). Woody cuttings of willows or other riparian species used in the structures will sprout (fig. 7.6.10D) and provide rapid revegetation (Hoag and Landis 2001).

Figure 7.6.10—*Restoration sites require special and sometimes extreme preparation before they can be planted. The blast zone of Mount St. Helens in Washington State was covered with volcanic ash (A), which had to be dug away so that seedlings could be planted in mineral soil (B). Stream banks often require bioengineering structures (C) for stabilization; when willow cuttings are used, they can sprout quickly (D) (D, courtesty of Steinfeld and others 2008).*

7.6.5 Selecting Plant Spacing and Pattern

The pattern and spacing of outplanted plants is also a reflection of project objectives. Industrial forestry projects, where timber production is the primary objective, dictate a specific number of seedlings per area in a regularly spaced pattern (fig. 7.6.11A) based on expected survival rates and laws governing the number of free-to-grow plants required after a specified period of time. Most planting projects will specify a certain desired number of established plants per area (table 7.6.2). These density targets should be considered general guidelines and should never override the selection of planting spots in biologically desirable areas (Paterson and others 2001).

Where ecological restoration is the objective, however, random outplanting of individual plants (fig. 7.6.11B) or outplanting in random groups (fig. 7.6.11C) is more representative of natural vegetation patterns.

The best place to plant nursery stock depends greatly on site conditions. When reforesting a level farm field with relatively uniform terrain, proper spacing is of utmost importance to minimize competition after the seedlings reach pole size. In this situation, then, the choice of planting spots is very mechanical; planters work in parallel lines and plant at the prescribed distance between spots (table 7.6.2). The same goes for using mechanized planters that plant seedlings at regular intervals.

7.6.5.1 Selecting planting spots

Microsites. When hand-planting in mountainous sites with old stumps and other woody debris, choosing the best planting spots is critical and more important than exact spacing. Planting in favorable microsites protects nursery stock and greatly improves the probability of survival. Examples of unfavorable planting spots include depressions with standing water, rocky spots, deep duff, and compacted soils. Seedlings shaded by a stump, log, or large rock tend to grow well, especially on hot, dry sites (fig. 7.6.12A&B). High sunlight on plant foliage causes moisture stress, and direct sunlight can cause lethal temperatures to the seedling stem at the ground line. Planting around physical obstructions also provides protection from cattle damage and large-game browsing (USDA Forest Service 2002). In the southern Rocky Mountains, planting in microsites shaded by dead woody material doubled the survival of ponderosa pine seedlings. This improved performance was attributed to better moisture and temperature and protection against animal browsing (Nelson 1984).

Where planting sites have been mechanically prepared with disc scarifiers, nursery stock should be planted on the side of the hole in mineral soil (fig. 7.6.12C). On mounds, the best planting spot is on the top (fig. 7.6.12D). Advance planning, crew training, and good supervision are essential to achieving good outplanting success.

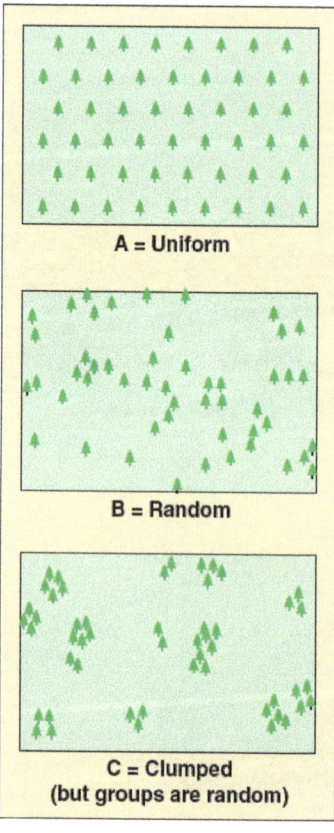

Figure 7.6.11—*In addition to target plant specifications, the objectives of the outplanting project affect planting patterns. If the objective is rapid growth or Christmas trees, then plants can be regularly spaced (A). Most restoration projects do not want the "cornfield look," however, so plants are spaced in a more random pattern to mimic natural conditions (B). The most natural outplanting look uses the random clumped pattern, where different species are planted in groups (C).*

Table 7.6.2—*Plant spacing based on regular grids with resultant stocking densities (modified from Cleary and others 1978)*

Spacing (m)	Plants per hectare	Plants per acre	Spacing (ft)
6.4 by 6.4	247	100	20.9 by 20.9
14.8 by 14.8	494	200	4.5 by 4.5
3.7 by 3.7	741	300	12.0 by 12.0
3.2 by 3.2	988	400	10.4 by 10.4
2.8 by 2.8	1,236	500	9.3 by 9.3
2.6 by 2.6	1,483	600	8.5 by 8.5
2.4 by 2.4	1,730	700	7.9 by 7.9
2.2 by 2.2	1,977	800	7.4 by 7.4
2.1 by 2.1	2,224	900	7.0 by 7.0
2.0 by 2.0	2,471	1,000	6.6 by 6.6

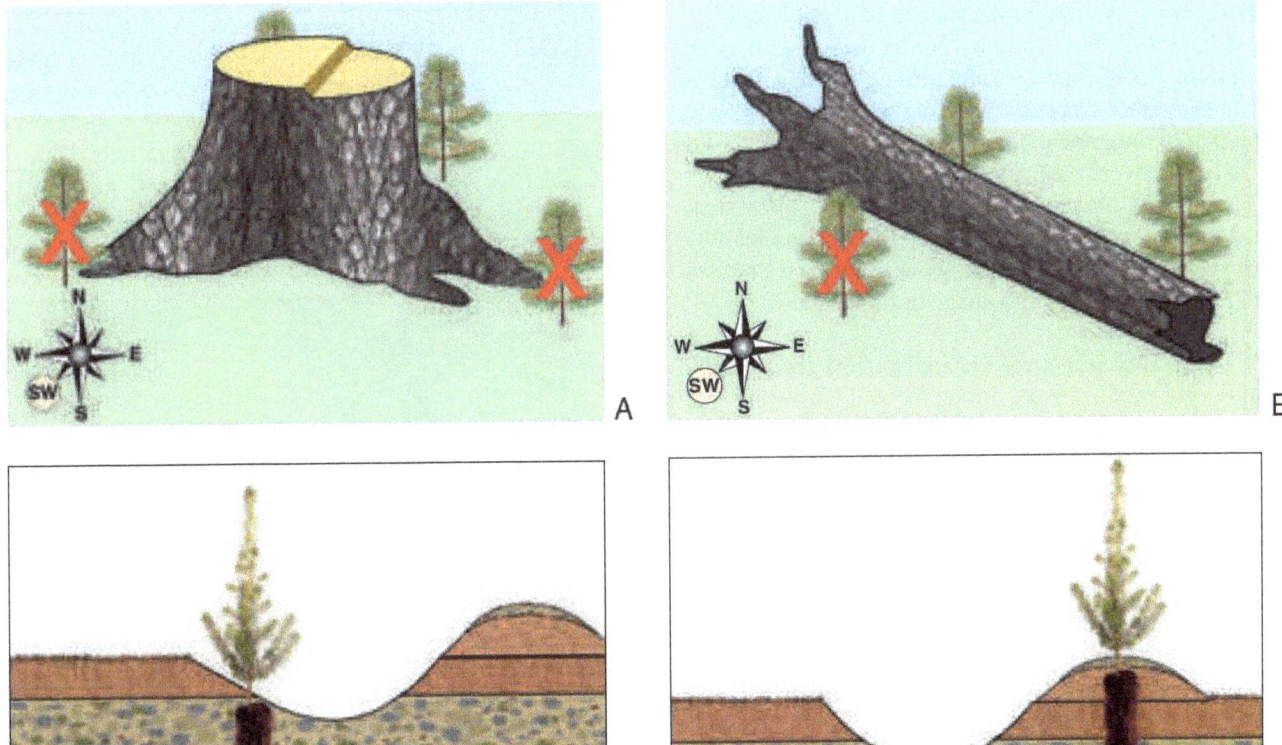

Figure 7.6.12—*On sites with uneven terrain or physical obstructions, the best planting spots are in microsites in the shade of stumps (A) or other debris (B). Specific planting spots are also prescribed where sites have been prepared by discing (C) or mounding (D) (A&B, from Rose and Haase 2006; C&D, from Heiskanen and Viiri 2005).*

7.6.6 Crew Training and Supervision

7.6.6.1 Plant handling

During the planting process, plants should always be handled with extreme care. Crews should be instructed never to toss or drop boxes of plants from the truck. Research shows that dropping seedlings from various heights can result in growth reductions after outplanting (fig. 7.5.5) (McKay and others 1993; Sharpe and others 1990; Tabbush 1986). Planters should never shake or beat plants to remove excess media. Deans and others (1990) found that height growth of Sitka spruce (*Picea sitchensis*) seedlings was negatively affected by beating them against boots at time of outplanting.

Each planter should carry only as many plants as can be planted in an hour or two. On larger reforestation and restoration projects, it is most efficient to use runners that carry batches of nursery stock on all-terrain vehicles (ATVs) from onsite storage to planters (fig. 7.6.13A). Planting bags must not be overfilled to avoid crushing plants (fig. 7.6.13B); loose plants are easier to remove without damage. After a planting hole has been prepared, only one plant should be pulled gently from the bag to avoid root stripping and stem damage (fig. 7.6.13C&D). One mistake inexperienced planters make is to take a handful of plants from the bag and then carry them from one planting hole to the next, increasing the risk of physical damage or desiccation.

The critical concepts are to handle plants gently and to minimize root exposure during the entire planting process. Although it is difficult to actually measure stresses during the shipping, handling, and outplanting process, comparison of outplanting performance between operational projects and research trials proves that the additional care afforded plants in trials pays off.

7.6.6.2 Proper planting technique

The retention of experienced tree planters from year to year appears to be declining (Betts 2008). Moreover, planting crews are often prone to high turnover rates during planting season, with members of the crew changing week to week. Nonetheless, it is crucial that all planters be thoroughly trained in planting procedures. Even a plant with the best quality will die if improperly outplanted. Good training, close supervision,

Figure 7.6.13—*All-terrain vehicles are handy for ferrying boxes of plants from onsite storage to planters (A). Planting bags should never be overfilled (B), and plants should be carefully removed one at a time from the container (C) or planting bag (D) and only after the planting hole has been dug (A, courtesy of Risto Rikala; B, courtesy of Mark Hainds; C, courtesy of J.D. Irving, Ltd).*

and regular inspection are important in order to optimize outplanting quality.

Somewhat surprisingly, very little is published on proper planting depths for container plants, although the advantages of "deep planting" bareroot stock are many (Stroempl 1990):

1. Improved stem stability against wind and snow pressure.

2. Insurance against root exposure from soil settling or washing away.

3. Protection of the "root collar" against heat injury.

4. Roots situated deeper in the soil profile have better access to soil moisture.

Therefore, on appropriate sites, the planting hole should be deep enough to bury the plug 2.5 to 5 cm (1 to 2 in)— about up to the cotyledon scar (Londo and Dicke 2006; USDA Forest Service 2002). This can vary with plant species; for example, in the Southeastern United States, longleaf pine seedlings that have their apical buds near the plug surface are planted with 0.6 to 1.3 cm (0.25 to 0.50 in) of the container plug exposed (Hainds 2003); this probably holds true for other species that have apical meristems near the plug surface. Because new photosynthates are required for new root growth after outplanting (van den Driessche 1987), burying foliage should probably be avoided.

The most important training concept is that good root-to-soil contact is necessary before nursery stock can become established on the site and quickly access water and mineral nutrients. The planting hole should be made deep enough so that, for most species, the root plug can be completely covered with mineral soil (fig. 7.6.14A) and "J-rooting" and unnecessary exposure of the root plug are avoided (fig. 7.6.14B), but the plug is not planted too deep (fig. 7.6.14C). According to Forest Service specifications, the minimum-size hole for container stock is 2.5 cm (1 in) deeper than plug length, and at least 7 cm (3 in) wider than the plug at top of the hole and 2 cm (1 in) at bottom (USDA Forest Service 2002). The planters should be instructed to plant at the correct depth and not to pull up on the plant to adjust depth or straightness. Plants should not be oriented more than 30 degrees from the vertical plane (fig. 7.6.14D); this seems obvious on level ground, but the steeper the slope, the more important this becomes. Planting holes should be backfilled with mineral soil without grass, sticks, rocks, or snow (fig. 7.6.14E). It is important to firmly tamp the soil around the root plug to remove air pockets (fig. 7.6.14F), but refrain from stomping around plants to avoid excessive soil compaction or stem injury.

Figure 7.6.14—*Nursery stock should be planted properly (A). Common problems include planting too shallow (B), planting too deep (C), improper vertical placement (D), filling the hole with debris (E), or poor compaction that leaves air pockets around the root plug (F) (modified from Rose and Haase 2006).*

Crew training is particularly important with volunteers or other inexperienced planters. Many of these people lack the skill or strength necessary to properly plant on wildland sites. One option is to have a professional create the planting holes with a machine auger and let the volunteers place and tamp plants into position. This technique has several benefits: the professional chooses the proper planting spot, creates the desired pattern, and makes certain that the planting hole is large and deep enough so that plants can be situated without "J-roots." Several studies have found that mechanical outplanting is more successful when working with private landowners who may not plant nursery stock properly (Davis and others 2004).

Although the choice of the proper planting tool is important, experienced planters can achieve success with a variety of implements. Planting failures are often more attributable to improper technique or handling rather than choice of planting tool (Adams and Patterson 2004).

Tree planting is strenuous work, and the swinging, bending, and lifting can quickly lead to worker injuries, especially early in the season. Back problems and carpal tunnel syndrome are common complaints. Crews should have sturdy boots, safety glasses, and hard hats and do strengthening and stretching exercises each day before starting to plant. The time and resources spent on worker protection will be offset by potential downtime and worker's compensation claims (Kloetzel 2004).

7.6.7 Hand-Planting Equipment

Root plugs on nursery stock used for reforestation or restoration are longer and narrower than plant materials used for landscaping and gardening, so specialty tools are necessary. Appropriate planting tools and technique can mean the difference between a live or dead plant, and between an on-budget or over-budget project (Kloetzel 2004).

Hand-planting methods provide maximum flexibility in plant placement and distribution. A well-trained and experienced hand-planter can surpass the planting quality and generally match the speed of many automated methods, especially over rough terrain. Hand-planting is especially recommended for placing plants into microsites and when planting a mixture of species or stocktypes. The most common types of hand-planting equipment are discussed in the following sections, but new equipment is continually being developed (Trent 1999).

7.6.7.1 Dibbles

Dibbles or dibble sticks were among the first tools used to plant container stock, primarily because they are easy to use (fig. 7.6.15A). Dibbles are custom-made probes that create a planting hole specific to one container type and size. Most designs have one or two metal foot pedals for forcing the point into the soil (fig. 7.6.15BA). After making the hole, the planter simply inserts the container plant and moves to the next hole. One drawback is the lack of loose soil to cover the top of the plug and prevent possible desiccation of the medium. Dibbles are most appropriate for lighter textured upland soils or alluvial bottomland soils in wetland restoration projects. Dibbles should be avoided on heavier textured clay soils, because they can compact soil and form a glaze around the planting hole that can restrict root egress (fig. 7.6.15C).

Hollow dibbles are a more recent modification that extract a core of soil and, therefore, do not cause soil compaction (fig. 7.6.15D). The hollow dibble heads are interchangeable, allowing use for different container sizes (Trent 1999). A slide hammer soil extractor can also remove a core of soil and, although one study found that it was more effective on rockier and compacted soils, it was also considered very strenuous to use because of its weight (Trent 1999).

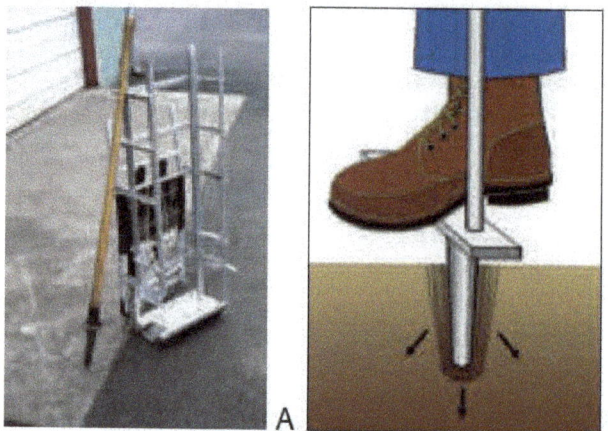

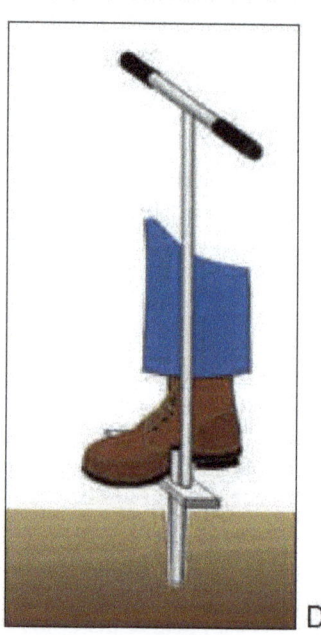

Figure 7.6.15—*Dibbles were among the first hand-planting tools developed for container nursery stock (A). Because they displace soil to form the planting hole (B), compaction can be severe enough to restrict root egress (C). Hollow dibbles are an improvement because they remove a core of soil to create a planting hole (D).*

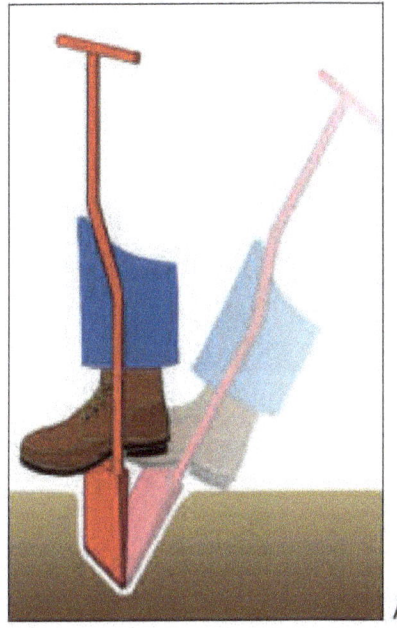  

Figure 7.6.16—Bars are easy-to-use planting tools that create a planting hole by lateral movement (A). The plant is positioned along one side of the planting hole (B), and soil is backfilled by leverage from the other side (C). Soil should be gently compacted around the plant with hand or foot (D).

Commercially produced dibbles are available for specific container types and sizes, including Ray Leach Conetainers™ and several cavity sizes of Styrofoam™ block containers (Kloetzel 2004). Dibbles have been used on shallow soils in Ontario but not on sites prone to frost heaving (Paterson and others 2001).

7.6.7.2 Bars

Planting bars originated with bareroot stock and are still used for smaller container plants. Bars are typically fabricated from a cylindrical bar with a wedge-shaped blade welded on the tip and side pedals to help force the blade into the soil. Like dibbles, planting bars require little experience or training. The bar is dropped and forced into the ground with the side pedals (fig. 7.6.16A), and the planting hole is formed by working the bar back and forth. The nursery plant is positioned vertically along one cut face (fig. 7.6.16B), and then the hole is closed by reinserting the bar into the soil on the opposite side of the planting hole and rocking the bar back and forth (fig. 7.6.16C). The final step is tamping any loose soil around the plant with the fist or boot to remove any air pockets (fig. 7.6.16D). In the Pacific Northwest, planting bars are often preferred for rocky soils but should not be used in heavier textured clays, where they cause excessive compaction (Cleary and others 1978). They also are popular on reforestation sites with sandy soils in the Southeastern United States. Planting bars are durable and simple to maintain, with only occasional blade sharpening required (Kloetzel 2004).

7.6.7.3 Hoedads

Hoedads, also known as planting hoes or mattocks, were developed specifically for planting bareroot conifer seedlings for reforestation and have since been adapted for container applications (fig. 7.6.17A). They are probably the most common handtool used in the United States, especially in the Pacific Northwest (Lowman 1999). Hoedads come in a variety of sizes and shapes and are one of the most versatile tools available. Special "plug hoes" for various sizes of container stock are available. Brackets, holding the hickory handle to the desired blade, are typically brass for extra weight and penetration, or tin alloy ("Tinselite") for lighter applications. Brackets can be found in two blade angle configurations: 100° angle for applications on gently sloped or flat areas and 90° angle for steep-ground planting. It is a good idea to purchase and keep handy spare blades, handles, and nuts and bolts with matching socket or box wrenches. Blades should be regularly sharpened with a metal file or electric grinding wheel (Kloetzel 2004).

Hoedads are particularly useful on steep reforestation sites, and even on rocky and compacted restoration projects. They are swung much like a pick, and it may take several swings to create a proper planting hole. With each swing, the planter lifts up and back with the butt of the handle to open the planting hole (fig. 7.6.17B). After a proper hole is complete, the planter uses the tip of the hoedad to gently loosen soil on the sides of the planting hole in order to avoid any compaction effects. Then, the plant is inserted and positioned to the proper depth (fig. 7.6.17C). While holding the plant, the planter used the hoedad blade to backfill the soil around the plug (fig. 7.6.17D). Finally, the planter gently tamps the soil around the plant (fig. 7.6.17E) and moves to the next planting spot. If plant competition is a problem, or if a planting basin is required, the back and side of the planting blade is a useful scalping tool (fig. 7.6.7C). Some compaction in the planting hole can occur on the backside of the blade with this tool, but compaction is typically less than with other methods.

Planting rates vary with container size, planter's skill, and terrain. Kloetzel (2004) reported that beginning planters can install 20 plants/hr while experienced planters may reach up to 100 plants/hr; on wetland planting projects

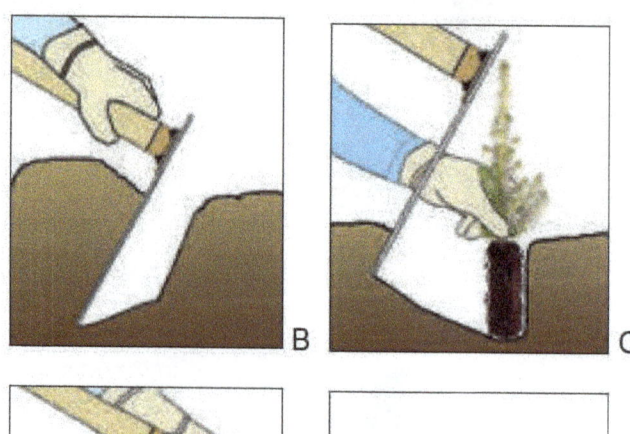

Figure 7.6.17—*Hoedads are one of the most popular planting tools in the mountains of the Western United States and Western Canada (A). After several swings to create a deep enough planting hole (B), the plant is positioned and held (C) while backfilling with soil (D). The final step is to gently compact the soil around the plant to remove any air pockets (E).*

with small stock and favorable soil conditions, production reached 240 plants/hr. For small-volume containers (66 cm^3 [4 in^3]), Meikle (2008) reports planting rates of 600 to 800 trees and shrubs per day on mineland reclamation sites, but the rate dropped to 400 to 600 plants when container volume increased to 164 cm^3 (10 in^3). Adding Vexar tubes to prevent herbivory dropped the planting rate by one-half (Meikle 2008).

7.6.7.4 Shovels

Although standard garden tile spades can be used, professional planters use customized shovels (fig. 7.6.18A) with blades long enough to accommodate large containers (fig. 7.6.18B). Wooden handles are standard, but fiberglass models are lighter, and reinforced blades (fig. 7.6.18C) can endure the vigorous prying action used to open planting holes (fig. 7.6.18D). Although shovels are not as difficult to learn to use as hoedads, planters should be trained to use tree-planting shovels efficiently. After the hole is excavated to the proper size and depth, the nursery plant is installed and held in a vertical position (fig. 7.6.18E) while the planter backfills around the root plug (fig. 7.6.18F). Tree-planting shovels are the tools of choice for some tree planters in the Western United States and are considered the most versatile planting tool in British Columbia (Mitchell and others 1990), as well as by reforestation crews in the Southeastern United States. Soil amendments, fertilizers, and other such in-soil treatments are easily installed with planting shovels. Sites requiring scalping require a two-person team with the scalper preparing the site beforehand. When using planting shovels, keep some spare handles and footpads on hand, along with tools for installing parts and sharpening blades (Kloetzel 2004).

In Ontario, experienced planters started the season by planting approximately 1,800 seedlings (100 cm^3 [6 in^3]) per day with shovels, while rookie planters managed only about 900. After about 6 weeks of planting, however, both groups were able to plant substantially more plants: 2,500 per day for experienced planters and 1,800 for rookies (Colombo 2008). In Washington State, larger stocktypes (250 cm^3 [15 in^3]) are planted west of the Cascade Mountains, and this is reflected in the planting rate. Only 900 larger seedlings can be planted per day compared to 1,000 of the smaller stocktype (Khadduri 2008).

7.6.7.5 Tubes

Planting tubes are mechanized dibbles that create a planting hole by compressing soil to the sides and bottom with a pointed pair of hinged jaws (fig. 7.6.19A). The jaws are switched open with a foot lever, and a container plant is dropped though the hollow stem tube into the hole (fig. 7.6.19B). The Pottiputki planting tube is the most popular brand and is available in several models with different tube diameters. In some models, the planting depth is adjustable, which would be necessary for stocktypes with longer plugs. One attractive benefit of planting tubes is less worker fatigue because the operator does not have to bend over. Planting tubes are popular in the Northeastern United States and Eastern Canada. Although popular in Ontario, they are considered expensive to purchase and maintain (Paterson and others 2001). In one comparison, planting tubes were just as effective as dibbles or planting bars (Jones and Alm 1989).

7.6.7.6 Motorized Augers

Power augers have been used in reforestation for decades and are becoming popular for restoration projects (fig. 7.6.20A). Augers work best in deep soils without too many large rocks or roots and are the best planting tool to use for larger, taller stocktypes. One concern has been compaction or glazing on the sides of the auger holes under some soil conditions (Lowman 1999), but this can be minimized by rocking the bit slightly. In Quebec and Nova Scotia, large-container seedlings are preferred because of heavy brush competition and a gasoline-powered auger was considered a better planting tool than spades or soil extractors in all soil types (St-Amour 1998). A gasoline-powered hand drill can be used with auger bits from 2.5 to 10 cm (1 to 4 in) in diameter, and the reversible transmission helps if the bit becomes stuck (Trent 1999).

One benefit of auger-planting projects is that the operator selects the location of planting spots and also controls the quality of the planting holes (fig. 7.6 20B). One operator can drill enough holes for several planters to follow and plant the nursery stock (fig. 7.6.20C). When scalping is required, the scalper will select the planting spots and create the scalp in advance of the auger operator. In some soil types, the operator will have to excavate extra mineral soil near each hole to ensure proper planting. Digging

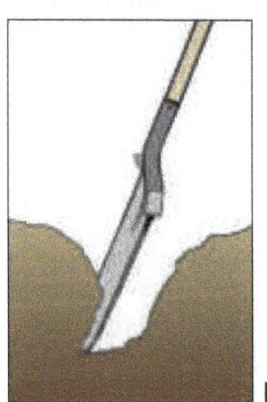

Figure 7.6.18—*Shovels are very versatile planting tools (A) and are ideal for large and deep container plants (B). Specialized shovels have reinforced blades (C) that open deep planting holes without soil compaction (D). While holding the plant vertical in the middle of the hole (E), firm the soil around the root plug as the hole is backfilled (F).*

Figure 7.6.19—*Planting tubes have pointed jaws that open the planting hole (A). The plant is dropped down into the hole through the hollow stem (B).*

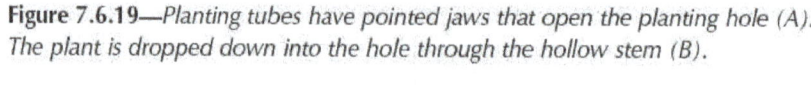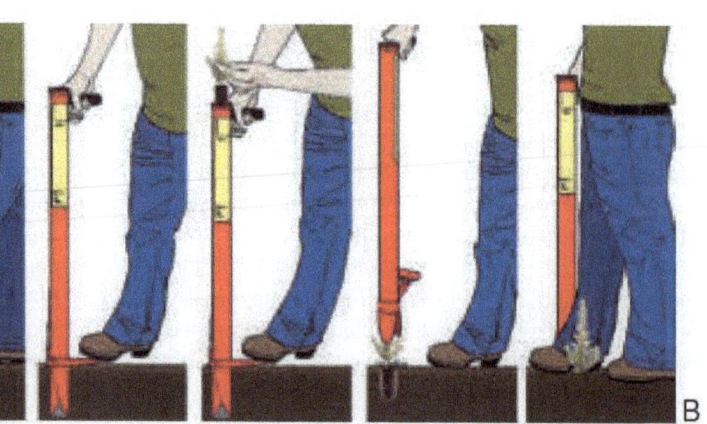

auger holes deeper than the depth of the container plug reduces compaction and can promote downward root growth. This means that the planter has to support the plant at the proper depth in the hole, while filling with soil from the bottom up (fig. 7.6.20D). Soil settling can be a problem with auger planting so it is a good idea to mound soil around the base of the plant. When more than one auger operator is available, it is best for them to take turns in order to reduce fatigue (Cleary and others 1978).

A wide variety of augers are commercially available for rent or sale: chainsaw-head, one-person, two-person, and tractor-mounted augers (fig. 7.6.20E). Most small planting projects can rent power augers at any commercial rental agency. When doing large scale reforestation or restoration projects, it is more cost effective to purchase one. If you are inexperienced with their operation, however, it is probably a good idea to rent first to make sure that you have the correct machine for the project. Augers are high-maintenance planting tools so keep a spare one available as well as extra parts and bits (Kloetzel 2004).

Well-organized auger teams can reach production rates ranging from 30 to 70 plants per person/hr (Kloetzel 2004). In Hawai'i, using an auger has become the ideal planting tool when volunteers or other non-professional planters are involved because the planting rate in order is 2.5 times that of standard hand tools (Jeffrey and Horiuchi 2003).

A

B

C

D

E

Figure 7.6.20—*Augers are effective planting tools because one skilled operator can create planting holes (A&B) and several other workers plant the stock (C) and fill the holes by hand (D). Tractor-mounted augers can create holes large enough for the biggest container stock (E).*

7.6.8 Machine-Planting

Planting machines have been used for forest tree seedlings for more than 100 years, and container plants are ideal for machine-planting because of their compact root systems and uniformity. Steadily increasing labor costs and difficulty in finding skilled planters has motivated many reforestation and restoration specialists to look at machine-planting (Hallonborg 1997). Foresters in British Columbia conducted trials with planting machines and found that the cost of machine-planting was comparable to hand-planting but was possible only on relatively flat, easily accessible sites. Many mountainous reforestation sites are steep, rocky, and covered with stumps and slash, factors that favor trained hand-planters (Mitchell and others 1990). Likewise, mechanical tree-planters are not used widely in Ontario because of site restrictions, high initial investment, and greater maintenance costs (Paterson and others 2001). Machine-planting has been more popular on the more gentle terrain of the Central, Northeastern, and Southern United States, and in Scandinavia.

Two basic types of planting machines, towed and self-propelled, are used and will be discussed separately (table 7.6.3).

7.6.8.1 Machines towed behind tractors

Many mechanical tree-planters are commercially available and consist of a rolling coulter, a trencher, an operator's seat, and packing wheels that are mounted on a sturdy frame (fig. 7.6.21A). Machine-planters for open fields feature a three-point hitch and are towed behind a tractor with the operator facing forward. The coulter cuts through the sod and any roots, and the trencher opens a narrow furrow (fig. 7.6.21B) in which the stock is hand-planted by the operator. Packing wheels on the back of the machine close the furrow and firm the soil around each plant. For planting in open fields, planting machines can also be equipped with a tank for applying herbicides (fig. 7.6.21C).

Some models, such as the Whitfield Tree Planting Machine, are popular for reforestation sites that have a lot of woody slash. They are safer because the operator rides backward in a protective cage and cannot be hit by debris thrown up by the tractor (fig. 7.6.21D). The operator places plants in a clip on a revolving chain assembly (fig. 7.6.21E) that carries the plant around until it is positioned in the furrow. The clips open mechanically and the plant is placed in the

Table 7.6.3—*Characteristics of the two types of tree-planting machines (modified from Landis 1999)*

Type of propulsion	Planting method	Plant placement	Plant spacing	Planting stock characteristics		
				Plug length determined by	Firm root plugs required	Stem rigidity required
Towed behind tractor	Furrow with closing wheels	Manual	Fixed in row	Depth of opening shoe	No	Yes
Self-propelled: mounted on excavator or harvester	Scarifying, mounding, and hole-making heads	Automated: hydraulic or pneumatic	Variable	Depth of planting head	Yes	Yes

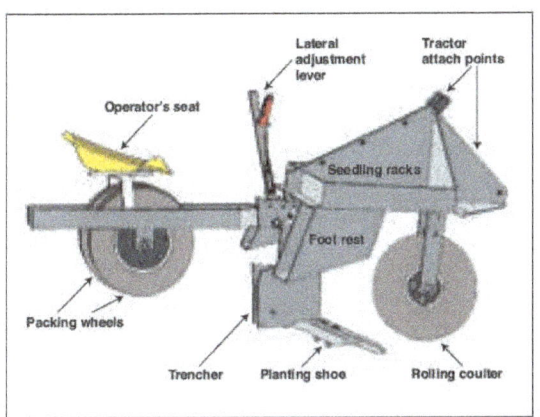

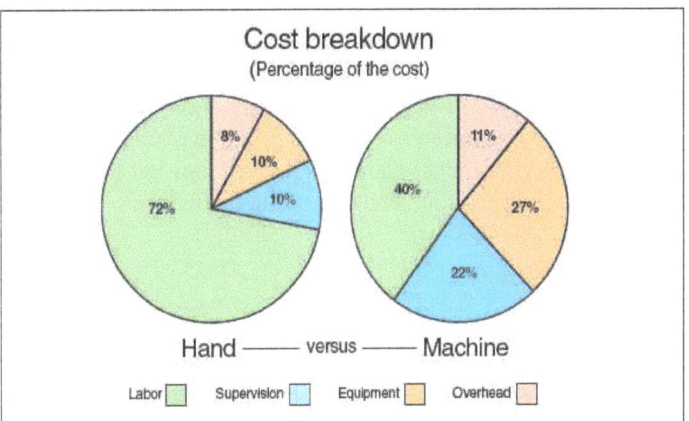

Figure 7.6.21—*A traditional type of planting machine (A) is towed behind a tractor, and the plants are spaced evenly in straight rows (B). Some models feature a herbicide sprayer for weed control (C). With the Whitfield planter, the operator rides backward and places seedlings in clips in a revolving chain (D) that carries them to the bottom of the furrow, which is closed by packing wheels (E). Economic comparisons have shown that machine-planting can be much more economical than hand-planting (F).*

furrow that is closed by the packing wheels (fig. 7.6.21E). The Taylor Tree Planting Machine is attached to the prime mover with a three-point hitch that allows down pressure to maintain planting depth; it also features a water tank on top to cool the stock (Converse 1999). Some machine-planters are equipped with furrowing attachments to scalp the planting spot, while others have spray attachments for applying herbicides to control unwanted vegetation. Planting speed varies with the ground conditions, size of the nursery stock, and experience and skill of the planting crew. Planting rates of 400 to 1,000 trees/hr have been reported (Slusher 1993), and in the Southeastern United States, 1,100 longleaf pine seedlings can be planted per hour at a within-row spacing of 4 m (12 ft) (South 2008).

Machine-planting must be evaluated on a site-by-site basis and is not effective on slopes greater than 35 percent. To offset the considerable transportation, operation, and maintenance costs, planting projects must be relatively large and accessible. A comparison of hand-planting versus machine-planting showed that labor savings can be considerable (fig. 7.6.21F). For example, in Southeastern Alaska, reforestation costs ranged from $247 to $321/ha ($100 to $130/ac), which was 18 percent less than hand-planting (Peterson and Charton 1999).

One consideration with towed planting machines is that plants are spaced regularly along the furrows. This is beneficial when a grid-like planting pattern is desired, such as in commercial forest or Christmas tree plantations (fig. 7.6.11A). Equal plant spacing is a drawback, however, when a more natural-appearing planting is desired (fig. 7.6.11B&C).

7.6.8.2 Self-propelled planting machines

Because of the high cost and unreliability of skilled planters, several models of self-propelled planters have been developed for container stock in Scandinavia (fig. 7.6.22A). These all-purpose planting machines have multiple benefits (Drake-Brockman 1998):

- Scarifying, mounding, and planting can be accomplished in a single operation.
- Planting spots are selected by the operator, which results in a more natural-looking plantation (fig. 7.6.11B&C).
- Fewer worker injuries as the machine does the physical work.
- Operators protected from inclement weather.
- Consistent quality of planting.
- Less contact with chemically-treated nursery stock.
- Reduced management planning and supervision.

Each planting machine has a different mechanism, but all have remote heads that scarify, mound, and plant seedlings in specific spots selected by the operator.

Bräcke planting machine. Developed in Sweden, this machine has been the most popular of the self-propelled planting machines (fig. 7.6.22A) and has been used in the United Kingdom and throughout Scandinavia. More than 30 Bräcke planting machines are being used in Finland due to a shortage of hand-planters. The quality of work has been equal to manual planting, but planting costs have been slightly higher (Harstela and others 2007). The planting head is mounted on the hydraulically controlled arm of an excavator or harvester (fig. 7.6.22B) and contains a circular magazine containing 60 to 88 plants (fig. 7.6.22C). It can create mounds and plant seedlings in the same operation (fig. 7.6.22D); production rates have varied from 140 to 250 seedlings/hr, depending on site conditions.

M-Planter. This Finnish planting machine is also mounted on a harvester or excavator boom, but it can create and plant two mounds without relocating (fig. 7.6.22E-F). The M-Planter features a larger seedling cassette that contains 242 seedlings and, in a recent comparison, outplanted the Bräcke by 24 to 38 percent over a variety of site conditions. Research is currently under way on an improved model of the M-Planter (Harstela and others 2007).

Ecoplanter. This Swedish planting machine is also mounted on a harvester or excavator boom, but it can create and plant two mounds at a time. The Ecoplanter has a capacity of 240 plants and can plant 220 to 250 seedlings/hr (Saarinen 2007).

Several comparisons of self-propelled planting machines were done in northern Europe. In Finland, the Bräcke and Ecoplanter had similar planting rates of 200 to 250 seedlings/hr. The planting quality of the Bräcke machine was comparable with hand-planting and better than the

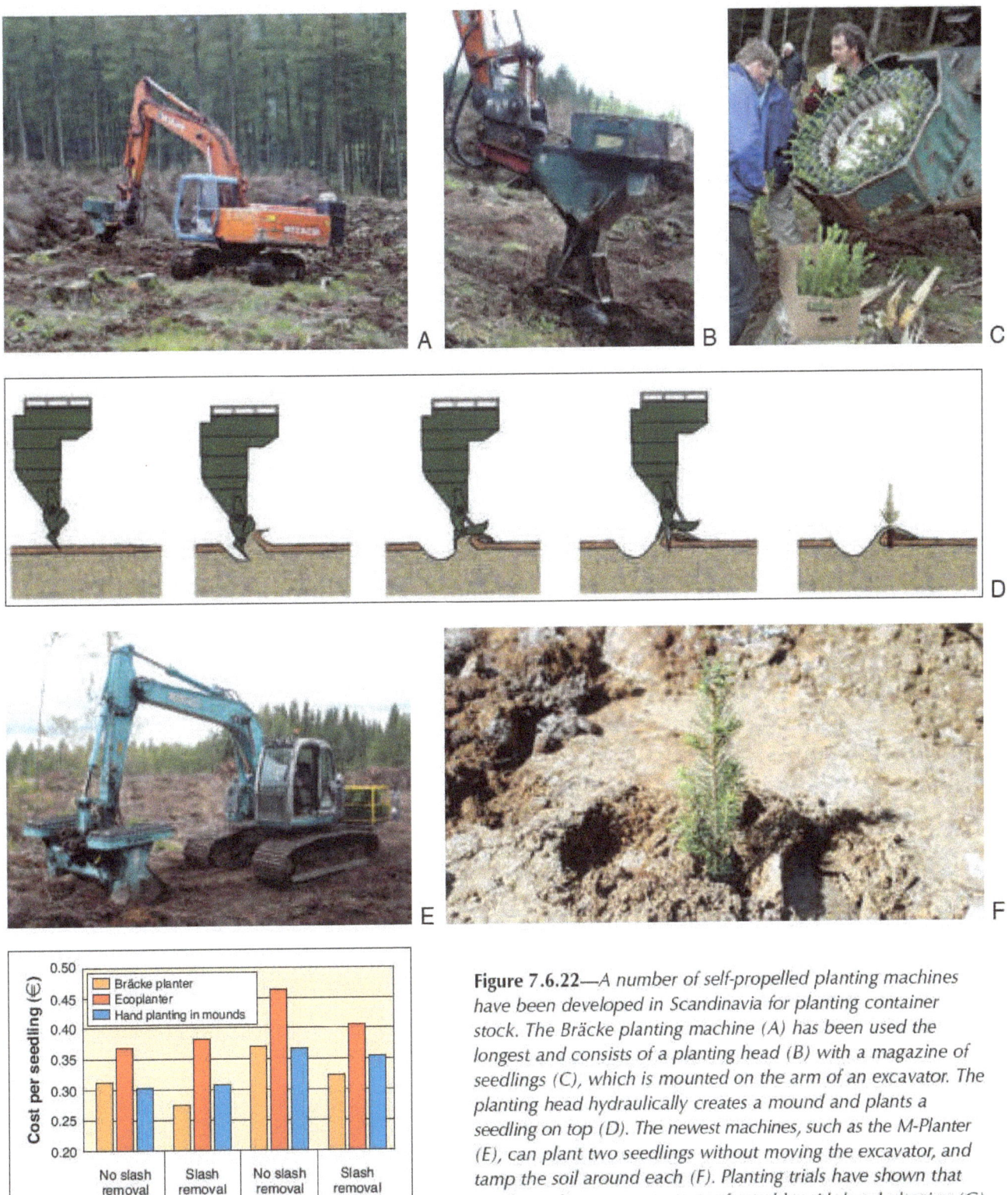

Figure 7.6.22—*A number of self-propelled planting machines have been developed in Scandinavia for planting container stock. The Bräcke planting machine (A) has been used the longest and consists of a planting head (B) with a magazine of seedlings (C), which is mounted on the arm of an excavator. The planting head hydraulically creates a mound and plants a seedling on top (D). The newest machines, such as the M-Planter (E), can plant two seedlings without moving the excavator, and tamp the soil around each (F). Planting trials have shown that machine-planters can compare favorably with hand-planting (G) (E, courtesy of Pekka Helenius; F, courtesy of Leo Tervo).*

Ecoplanter (fig. 7.6.22G), which caused stem deformation and had more weakened or dead trees after 2 years (Saarinen 2007). In a test of the Bräcke in Ireland, planting quality was well within planting quality specifications, but not as good as hand-planting. However, no overall significant differences in height growth and root-collar diameter increment were found after the first growing season (Nieuwenhuis and Egan 2002). In the United Kingdom, the Bräcke machine produced acceptable planting of container conifers on upland reforestation sites (Drake-Brockman 1998).

7.6.9 Planting Equipment for Large Stock

Large container stock and nonrooted cuttings are difficult to plant effectively, so special equipment has been developed. Note that good access is essential and, in the case of the pot planter, a source of water must be available.

7.6.9.1 Expandable stinger

The expandable stinger is a recently developed planting machine attached to the arm of an excavator (fig. 7.6.23A) that creates a hole and plants nursery stock in one operation. The planting head is composed of two parallel steel shafts, which are hinged in the middle to open and close in a scissor-like manner. Each shaft is constructed to create a long, hollow chamber between them when closed. The opening and closing of the shafts are hydraulically driven. When the shafts are closed, the stinger comes to a point and is pushed into the soil by the force of the excavator arm. A long hardwood cutting or container plant is placed into the chamber. The expandable stinger is maneuvered to the planting spot, where the beak is inserted into the soil. When the beak opens, the plant drops to the bottom of the hole (fig. 7.6.23B).

Two expandable stinger models, single-shot and 50-shot, are currently in use. The single-shot model holds only one plant at a time and averages 50 to 80 plants/hr. The rotary magazine of the 50-shot model holds 50 plants of up to three different species and can double the planting rate of the single-shot model (Kloetzel 2004). The expandable stinger can reach sites that are inaccessible by other planting equipment. The arms on smaller excavators can reach 7.5 m (25 ft), while those on larger machines extend out to a 15-m (50-ft) radius. This equipment can also plant in very rocky soil conditions, including rip-rap and gabions, and can penetrate very compacted soils making it ideal for restoration projects. It is a good idea to have someone follow the stinger and fill in around the plants with mineral soil.

The major drawback to the expandable stinger is its expense. In addition to hourly operating costs, mobilization costs can be very high, although these costs should be amortized across the entire project. As the number of plants installed by the expandable stinger on a project increases, the cost per established plant decreases. In a well-planned operation, the expandable stinger can achieve a production rate of 200 plants/hr.

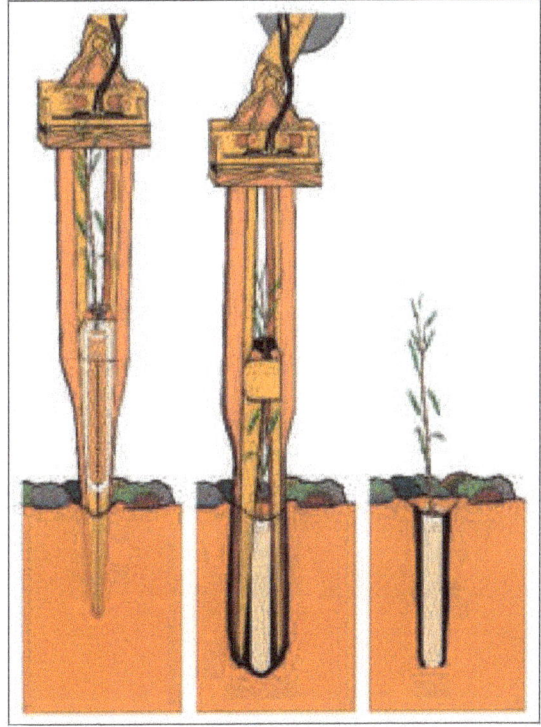

Figure 7.6.23—*The expandable stinger is a specialized planting machine for harsh sites, including compacted soil and rip-rap (A). The long scissor-like planting head creates a planting hole where a tall pot container or nonrooted cutting can be installed (B).*

7.6.9.2 Pot planter

The pot planter was developed for riparian restoration projects (Hoag 2006) and uses high-pressure water to create planting holes for large container stock. Water from a lake, stream, or tank is pumped into a compressor (fig. 7.6.24A) and then forced through the tip of a high pressure nozzle (fig. 7.6.24B). The pot planter has 7.6-cm (3-in) vanes attached to the sides of the nozzle, which create holes large enough for containers up to 3.8 L (1 gal) (fig. 7.6.24C). The hole that is created by the pot planter is filled with a soil slurry that is displaced when the root plug of the container plant is inserted to the desired planting depth. After the water drains from the slurry into the surrounding soil, the soil settles in around the root plug, assuring good soil-to-root contact. The water also thoroughly wets the root plugs and seeps into the surrounding soil. Operational trials have shown that large container stock can be planted at a rate of approximately 60 plants/hr (Hoag 2006).

Figure 7.6.24 —*The pot planter uses high-pressure water pumped from a compressor (A) through holes in a specialized nozzle (B) to create planting holes for large container stock (C).*

7.6.10 Treatments at Time of Planting

Depending on the site, several other treatments may be applied to plants at the time of outplanting to improve survival and growth. These solutions to potential limiting factors would have been identified during the site evaluation (see Section 7.6.1).

7.6.10.1 Protection from animal damage

Compared with wild plants, fertilized nursery stock has higher levels of mineral nutrients and is therefore preferred browse by deer, elk, and other animals (Fredrickson 2003). Plants (especially the terminal shoots) are eaten by deer, elk, gophers, and other animals, although the rate of browsing can vary by season (Kaye 2001) (fig. 7.6.25A). If the outplanting area is known to have a problem with animal damage, then control measures may be necessary. Physical barriers installed immediately after planting, such as netting, rigid mesh tubing (fig. 7.6.25B), bud capping, and fencing, can help protect plants long enough for them to grow large enough to resist animal damage. Troy and others (2006) found that 95 percent of nonprotected oak (*Quercus* spp.) seedlings were browsed, compared with only 4 percent of those in protective shelters. Johnson and Okula (2006) concluded that browse protection increased both survival and growth of outplanted antelope bitterbrush (*Purshia tridentata*) seedlings.

A variety of solid-walled and mesh tree shelters is available, and the environment and plant response can vary considerably. Western redcedar and Oregon white oak (*Quercus garryana*) container stock was outplanted in fine-mesh fabric shelters; solid-walled white shelters with and without vent holes; and solid-walled, blue, nonvented shelters. One year after outplanting, height and diameter growth of the western redcedar were significantly increased in all shelter types, with blue solid-walled shelters resulting in the greatest height growth. In blue, solid-walled shelters, however, photosynthesis and stem diameter growth of Oregon white oak seedlings, which are less tolerant of shade, were significantly reduced compared with nonsheltered seedlings (Devine and Harrington 2008).

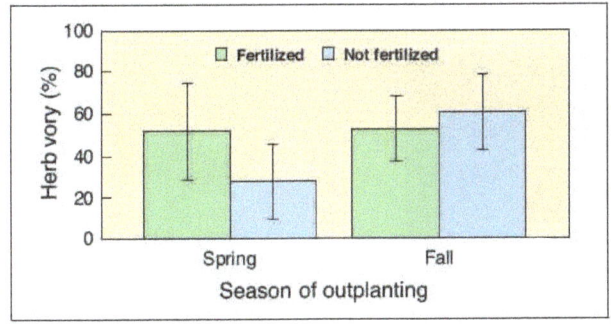

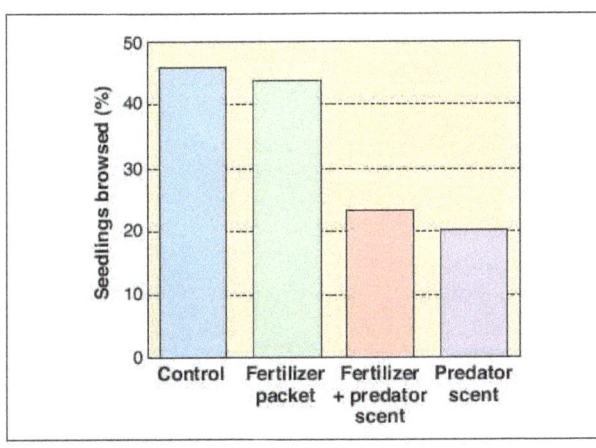

Figure 7.6.25—*Browsing damage to outplanted seedlings can be very high on some sites (A). Options for protecting plants from animal damage include plastic mesh tubing (B), or an application of predator scent repellent (C).*

Chemical repellents are another option to protect from animal damage. These repellents are less costly than physical barriers but their efficacy can be shorter lived. A variety of products that have an odor or taste that is repugnant to wildlife is available. Treating plants with these products can result in significant reductions in browsing (Frank 1992; MacGowan and others 2004) (fig. 7.6.25C).

7.6.10.2 Fertilization

Mineral nutrition is a key component of plant performance after outplanting, and most outplanting sites are limiting for many essential nutrients, especially nitrogen. Agriform® tablets are made of ureaformaldehyde, which provides slow release of nitrogen, as well as phosphorus, potassium, and other secondary and minor elements (Scotts 2007). Although commonly used in ornamental plantings, Agriform® tablets have yet to find wide use in forestry, conservation, or native plant outplantings.

Instead, polymer-coated controlled-release fertilizers (Osmocote®, Apex®, Multicote®, Nutricote®, Diffusion®) have become the most popular method of fertilizing at the time of outplanting (Jacobs and others 2003); they feature nutrient release rates up to 18 months. Fertilizer prills are incorporated into the growing medium during sowing (Moore and Fan 2002; Haase and others 2006) or added to the bottom of the planting hole (Arnott and Burdett 1988; van den Driessche 1988). Other applications include applying the fertilizer prills in a dibbled hole alongside the plant or broadcasting it around its base. To minimize the possibility of fertilizer burn to roots and to prevent the nutrients from being "stolen" by competing vegetation, the side application makes the most sense (Landis and Dumroese 2009).

Fertilizer efficacy, however, varies with site conditions (Rose and Ketchum 2002, Everett and others 2007). On moisture-limiting sites, fertilizer salts can buildup to toxic levels, resulting in a negative effect on survival and

Figure 7.6.26—*Mulching with paper mats (A&B) or loose materials (C) can reduce competing vegetation around the planted seedling.*

growth (Jacobs and others 2004). For fall outplanting in Northern California, the initial growth benefits of controlled-release fertilization did not hold up over time (Fredrickson 2003). Before applying any fertilizer, it is crucial to consider the formulation, application rate, placement, solubility/release rate, and existing nutrient levels on the site.

7.6.10.3 Mulches

In addition to site prep practices that minimize effects of competing vegetation (see Section 7.6.4.5), mulching can reduce recurrence of vegetative competition for a longer duration than initial site preparation. Mulch mats made from materials such as plastic, fabric, sod, or paper (fig. 7.6.26A&B) are held in place with rocks, branches, or stakes. Mulching plants can also be accomplished with a thick layer of loose organic matter, such as corn cobs, coconut fiber, pine straw, sawdust, or bark chips (fig. 7.6.26C). In addition to inhibiting growth of competing vegetation, mulch insulates soil from temperature extremes and helps maintain soil moisture by reducing surface evaporation. Although purchase and installation of mulch materials can be costly, mulches can significantly improve plant survival and growth on droughty sites. McDonald and others (1994) found that large (3 m x 3 m [9 ft x 9 ft]), long-lasting (5 years) mulch mats effectively enabled ponderosa pine seedlings to become established on the site unimpeded by competing plants and significantly increasing height and diameter growth compared with controls. Similarly, bur oak (*Quercus macrocarpa*) and white ash (*Fraxinus americana*) had a significant positive response to mulch treatments (Truax and Gagnon 1993). On dry restoration sites, mulches can be particularly effective. Plastic mulches of only 122 cm (48 in) in diameter significantly increased soil water content and subsequent growth of Oregon white oak container seedlings; even post-planting irrigation was effective only under mulches (Devine and others 2007).

7.6.10.4 Shelters

As mentioned previously, tree shelters (fig. 7.6.27A) can protect plants from animal damage. Another important benefit is that shelters limit the intensity of UV light and drying winds that cause damage by desiccation and sun scald (fig. 7.6.27B). Engelmann spruce (*Picea engelmannii*) seedling survival increased from 58 percent to more than 95 percent when shelters were installed (Jacobs and Steinbeck 2001). Tree shelters are available in a variety of sizes and colors (allowing varying amounts of solar radiation to penetrate), as well as with or without venting. Selection of a specific shelter should be made based on expected site conditions and the growth habit of the species. In a comparison of ventilated and nonventilated shelters, ventilation consistently reduced inside shelter

Figure 7.6.27—*Tree shelters (A) protect plants from animal damage and sunscald (B); shading is also effective against sun damage but must be installed on the southwestern side of the plant (C).*

temperatures by about 2.7 °C (5 °F) (Swistock and others 1999). Plants kept in tall, rigid shelters for a long period of time can become spindly (reduced stem diameter relative to height) and incapable of standing upright after shelter removal (Burger and others 1996). Management considerations for using tree shelters should include the costs of purchase, assembly, and installation as well as annual maintenance following winter snow pack that can crush the shelters and cause plant damage. Nevertheless the increased cost may be offset by increased survival, thereby reducing the need to replant at a later date when competing vegetation is established.

7.6.10.5 Shading

Ideally, an outplanting site provides adequate materials, such as stumps or logs, to provide microsites for planting (see Section 7.6.5.1). It is sometimes useful, however, to install artificial shade to protect plants from damaging heat. Resistance to heat damage increases with plant size because the ability of the plant to shade itself increases. Heat damage usually occurs on flat or south-facing sites in regions with hot, dry summers and clear skies, but it can also occur in wetter regions under dry, clear conditions (fig. 7.6.27B). Shading only the basal portion of the stem appears to be as effective in preventing heat damage as shading the entire stem and some foliage, which can also reduce transpiration (Helgerson 1989a). Five-year survival of Douglas-fir seedlings was increased with shading on two south-facing sites in Southwest Oregon (Helgerson 1989b). In another study, artificial shading significantly increased seedling survival on four of six harsh sites west of the Cascade Mountains (Peterson 1982). Artificial shade materials, which include cardboard, shingles, rigid shade cloth, and other materials should be installed on the south or southwest side of the seedling (fig. 7.6.27C).

7.6.10.6 Irrigation

Although irrigation is impossible on typical reforestation sites, watering after outplanting is sometimes needed on severe restoration sites and special techniques are employed. For example, on a Sonoran Desert site, honey mesquite (*Prosopis glandulosa*) seedlings were irrigated through plastic pipes to ensure that the water reached the root zone without loss to evaporation. Four years later, plants that had been deep watered had three times better survival and were significantly taller than surface-watered plants. More information on deep watering and other irrigation techniques can be found in Bainbridge (2007) and Steinfeld and others (2007).

7.6.11 Monitoring Outplanting Performance

Reforestation and restoration outplantings are an expensive investment, so it makes sense to conduct surveys to evaluate their need, monitor performance, and track outplanting success over time. Many different types of reforestation surveys have been well covered in the literature (Pearce 1990; Stein 1992); and an excellent guide on how to evaluate restoration plantings can be found in chapter 12 of Steinfeld and others (2007).

The following discussion deals with monitoring planting quality during the project. The only way to determine if planting is being done correctly is to conduct an inspection right behind the planting crew (Neumann and Landis 1995). With contract planting jobs, these inspections certify whether the work meets specifications, and the results are used to calculate payment. Prompt and thorough inspections can also lead to increased outplanting success in subsequent projects. In Texas, for example, the incidence of plantation failure was more than cut in half (from 40 percent to about 16 percent) after an inspection program was initiated (Boggus 1994).

A typical plantation inspection consists of the following three steps (Rose 1992).

Check the number and spatial distribution of plants. Plots are established to determine whether the correct number of plants was installed in a given area, whether good planting spots were selected, and whether plants were properly spaced. New technology may make this job easier. In a recent research trial, a tree-planting dibble was outfitted with an accelerometer, a Global Positioning System (GPS) unit, and a data logger to map the locations of seedlings as they were planted. Results showed that the equipment accurately (±7 percent) counted the number of seedlings planted. Although the GPS system was not sensitive enough to identify individual plants, this may be resolved with increased sensitivity of newer equipment (McDonald and others 2008).

Aboveground inspection. A representative sample of plants is examined to see if the planting spot was selected properly and to check the quality of the scalping, stem orientation, planting depth, and use of natural or artificial shade. Planting depth, which is one of the most critical things to check, is usually specified in relation to top of the root plug (fig. 7.6.28A; see Section 7.6.6.2).

Belowground inspection. A hole is dug with a planting shovel (fig. 7.6.18C) alongside the planted plant to check for proper root orientation, loose soil, air pockets, foreign material in the hole, and so on. Begin digging the hole far enough away from the main stem (25 cm [10 in]) so that roots are not disturbed in the process of inserting the shovel. Then, gently clear soil away while digging toward the plug so that the plug can finally be inspected in the position it was planted (fig. 7.6.28B). The plug must be in a vertical plane and not twisted, compressed, or jammed and the hole should not contain large rocks, sticks, litter, cones, or other foreign debris. Soil should be nearly as firm as the undisturbed surrounding soil, with no air pockets. In auger plantings, be sure to check soil firmness near the bottom of the holes (USDA Forest Service 2002).

7.6.11.1 What type of survey is best?

Two types of surveys, circular plots and stake rows, traditionally have been performed, and each has its own advantages.

Circular plots. The traditional method for determining planting density is to measure 40-m^2 (1/100-acre) plots that are evenly distributed throughout the plantation. An adequate sample is about 2.5 plots per hectare (one plot per acre), with usually no more than 30 plots evenly distributed throughout the planted area. A 100th acre plot has a radius of 3.6 m (11 ft, 9.3 in), which is established with a center stake and a piece of string or twine cut to this length (Londo and Dicke 2006). Seedlings outplanted within the plot are counted, and their tops examined and measured. The root system of the plant closest to the center is excavated to evaluate planting technique. Record each plot separately on a survey form (fig. 7.6.28C) using the examination criteria shown in figure 7.6.14.

Stake rows. Rapid weed growth makes it surprisingly hard to locate desired plants, so 10-plant stake rows are used to make plants easier to find in subsequent evaluations. Establish a starting point that can be easily located and stake 10 plants along a compass bearing. Height, diameter, and plant condition are recorded on the data form, along with average spacing between plants. Stake row data are typically used to determine survival and growth rates and, with average spacing between plants, can also be used to calculate plants per area (Londo and Dicke 2006).

Figure 7.6.28—*It is best to inspect right after the planting crew (A). Dig a vertical hole alongside the seedling (B) to check for proper depth and alignment of the root plug. Using a standard survey form (C) will ensure that the same information is collected at each plot.*

Plantation:	Date Planted:	Inspector:
Plot Number:	Date Inspected:	Contract Number:

Plant No.	Species Code	Height (cm)	Caliper (mm)	Condition Codes	Comments	Plant Condition Codes
						1 = Poor planting spot
						2 = Planted too deep
						3 = Planted too shallow
						4 = "J" root
						5 = Poor compaction — Air pockets
						6 = Foreign material in hole
						7 = Not planted vertically
						8 = Poor scalp
						9 = Planted too close to another plant
						10 = Other — Provide comments
						Plot Map
						(concentric circle plot map, N indicated) Scale = _____

7.6.11.2 What sampling design is best?

Systematic stratified sampling is often recommended, because plots are located at standard predetermined distances and are therefore easy to establish and locate again at a later time. Stratification means that the entire population of plants in the outplanting area is subdivided into homogeneous units before sampling begins. First, strata of uniform conditions are identified, and then sample plots are located systematically within these areas (Pearce 1990). These strata could be based on species, nursery of origin, planting crew, or any other factor that could introduce serious variation. Machine-planted stock on abandoned farmland would have less variability because conditions are relatively uniform and planter-to-planter variation is not an issue. In contrast, considerable variability exists on hand-planted projects in mountainous terrain, where differences in aspect, soil, and planting technique occur (Neumann and Landis 1995).

7.6.11.3 How many plots are necessary?

The number of plots to establish is generally a function of two factors: (1) available resources (time and money); and (2) variability of the attributes that will be measured. In calculating an appropriate number of plots, statisticians are interested in some measure of variability, such as the standard deviation of plant heights in the outplanting. Using this example, if a quick check of height varies greatly within the plantation to be sampled, then more plots should be taken. On the other hand, if the heights appear to be very uniform, then fewer plots will be sufficient. If you want statistical significance, more complicated calculations are available to compute appropriate number of plots, using an estimate of the variability of the attribute and the degree of statistical accuracy desired (Stein 1992).

Determining the number of plots based on variability is often a judgment call but, in most cases, a 1- to 2-percent sampling intensity is sufficient (Neumann and Landis 1995).

7.6.12 Conclusions and Recommendations

Outplanting is the final stage in the nursery process, and survival and growth are the ultimate tests of plant quality. The final three steps of the Target Plant Concept are critical to outplanting success and should be considered when plannning and initiating outplanting projects. Each outplanting site is unique and should be evaluated to identify critical limiting factors as well as the best season for outplanting during the planning process. The best outplanting tool and technique must also be specified during planning, because that decision will have a major effect on the best stocktype to produce. A wide variety of hand- and machine-planting options are available, but each tool or technique is best suited to particular stocktypes and outplanting site conditions. All this information is traditionally included in the site prescription which will guide the entire nursery-to-outplanting process.

Stock-handling during transport and on the planting site has a critical effect on outplanting performance. Nursery stock should be outplanted as soon as it arrives, but often a day or two of onsite storage is necessary. It is wise to plan for contingencies, such as bad weather, crew problems, or equipment breakdown. A representative sample of the nursery stock should be inspected as soon as it arrives on the planting site to identify possible problems and make adjustments. At the same time, a survey of the planting site itself should be conducted and plans made for which areas should be planted first.

Site preparation treatments, which are also part of the site prescription, will ensure that the proper supplies and equipment are available ahead of time. Plant spacing and pattern should be specified in the prescription so this critical information is part of crew training. Other treatments, such as plastic netting, tree shelters, and mulches, may need to be applied to plants at the time of outplanting to ameliorate potentially limiting site factors.

The final step in the process is to conduct surveys during and immediately after outplanting to evaluate planting, monitor plant performance, and track outplanting success over time. The best type and intensity of sampling will depend on project objectives and should be designed as part of the site prescription. Successful outplanting projects are the result of good planning and timely execution. Often, adjustments need to be made onsite but most of these contingencies can be anticipated in the site prescription.

7.6.13 Literature Cited

Adams, J.C.; Patterson, W.B. 2004. Comparison of planting bar and hoedad planted seedlings for survival and growth in a controlled environment. In: Connor, K.F., ed. Proceedings of the 12th biennial southern silvicultural research conference. Gen. Tech. Rep. SRS-71. Asheville, NC: USDA Forest Service, Southern Research Station: 423-424.

Arnott, J.T.; Burdett, A.N. 1988. Early growth of planted western hemlock in relation to stock type and controlled-release fertilizer application. Canadian Journal of Forest Research 18: 710-717.

Bainbridge, D.A. 2007. A guide for desert and dryland restoration: a new hope for arid lands. Washington, DC: Island Press. 391 p.

Balisky, A.C.; Burton, P.J. 1997. Planted conifer seedling growth under two soil thermal regimes in high-elevation forest openings in interior British Columbia. New Forests 14: 63-82.

Barnard, E.L.; Dixon, W.N.; Ash, E.C.; Fraedrich, S.W.; Cordell, C.E. 1995. Scalping reduces impact of soil-borne pests and improves survival and growth of slash pine seedlings on converted agricultural croplands. Southern Journal of Applied Forestry 19: 49-59.

Bergsten, U.; Goulet, F.; Lundmark, T.; Lofvenius, M.O. 2001. Frost heaving in a boreal soil in relation to soil scarification and snow cover. Canadian Journal of Forest Research 31: 1084-1092.

Betts, J. 2008. Recent workforce trends and their effects on the silviculture program in British Columbia. In: Dumroese, R.K.; Riley, L.E., tech. coords. National Proceedings: Forest and Conservation Nursery Associations—2007. Proceedings RMRS-P-57. Fort Collins, CO: USDA Forest Service, Rocky Mountain Research Station: 164-165.

Boggus, T. 1994. Personal communication. Lubbock, TX: Texas State Forest Service.

Burger, D.W.; Forister, G.W.; Kiehl, P.A. 1996. Height, caliper growth and biomass response of ten shade tree species to tree shelters. Journal of Arboriculture 22: 161-166.

Burney, O.T.; Jacobs, D.F. 2009. Influence of sulfometuron methyl on conifer seedling root development. New Forests 37: 85-97.

Camm, E.L.; Guy, R.D.; Kubien, D.S.; Goetze, D.C.; Silim, S.N.; Burton, P.J. 1995. Physiological recovery of freezer-stored white and Engelmann spruce seedlings planted following different thawing regimes. New Forests 10: 55-77.

Campbell, B.; Kiiskila, S.; Philip, L.J.; Zwiazek, J.J.; Jones, M.D. 2006. Effects of forest floor planting and stock type on growth and root emergence of *Pinus contorta* seedlings in a cold northern cutblock. New Forests 32: 145-162.

Chavasse, C.G.R. (ed). 1981. Forest nursery and establishment practice in New Zealand. New Zealand Forest Service, Forest Research Institute, FRI Symposium No. 22, 591 p.

Cleary, B.D.; Greaves, R.D.; Hermann, R.K. 1978. Regenerating Oregon's forests. Corvallis, OR: Oregon State University, Extension Service. 286 p.

Colombo, S. 2008. Personal communication. Thunder Bay, ON, Canada: Ontario Forest Research Institute, Centre for Northern Forest Ecosystem Research.

Converse, C.M. 1999. Mechanical site preparation and tree planting equipment for Alaska. In: Alden, J. ed. Stocking standards and reforestation methods for Alaska. Misc. Pub. 99-8. Fairbanks, AK: University of Alaska, Agricultural and Forestry Experiment Station: 57-67.

Davis, A.S.; Jacobs, D.F.; Ross-Davis, A. 2004. Success of hardwood tree plantations in Indiana and implications for nursery managers. In: Riley, L.E.; Dumroese, R.K.; Landis, T.D, tech. coords. In: National proceedings, Forest and Conservation Nursery Associations -- 2003. Proceedings RMRS-P-33. Fort Collins, CO: USDA Forest Service, Rocky Mountain Research Station: 107-110.

Deans, J.D.; Lundberg, C.; Tabbush, P.M.; Cannell, M.G.R.; Sheppard, L.J.; Murray, M.B. 1990. The influence of desiccation, rough handling and cold storage on the quality and establishment of Sitka spruce planting stock. Forestry 63: 129–141.

Devine, W.D.; Harrington, C.A. 2008. Influence of four tree shelter types on microclimate and seedling performance of Oregon white oak and western redcedar. Res. Pap. PNW-RP-576. Portland, OR: USDA Forest Service, Pacific Northwest Research Station. 35 p.

Devine, W.D.; Harrington, C.A.; Leonard, L.P. 2007. Post-planting treatments increase growth of Oregon white oak (*Quercus garryana* Dougl. ex Hook.) seedlings. Restoration Ecology 15: 212-222.

Domisch, T.; Finér, L.; Lehto, T. 2001. Effects of soil temperature on biomass and carbohydrate allocation in Scots pine (*Pinus sylvestris*) seedlings at the beginning of the growing season. Tree Physiology 21: 465-472.

Drake-Brockman, G.R. 1998. Evaluation of the Bräcke Planter on UK restock sites. Tech. Note 7/98. Dumfries, United Kingdom: Forestry Commission, Technical Development Branch. 10 p.

Emmingham, W.H.; Cleary, B.C.; DeYoe, D.R. 2002. Seedling care and handling. In: The woodland workbook: forest protection. Corvallis, OR: Oregon State University Extension Service. 4 p.

Everett, K.T.; Hawkins, B.J.; Kiiskila, S. 2007. Growth and nutrient dynamics of Douglas-fir seedlings raised with exponential or conventional fertilization and planted with or without fertilizer. Canadian Journal of Forest Research 37: 2552-2562.

Fleming, R.L.; Black, T.A.; Eldridge, N.R. 1994. Effects of site preparation on root zone soil water regimes in high-elevation forest clearcuts. Forest Ecology and Management 68:173-188.

Frank, D. 1992. Predator odour as a deer browsing repellent: an investigation of an East Coast Vancouver Island Douglas-fir plantation. FRDA Research Memo No. 204.Victoria, BC, Canada: British Columbia Ministry of Forests.

Fredrickson, E. 2003. Fall planting in northern California. In: Riley, L.E.; Dumroese, R.K.; Landis, T.D., tech. coords. National Proceedings: Forest and Conservation Nursery Associations—2002. Proceedings RMRS-P-28. USDA Forest Service, Rocky Mountain Research Station: 159-161.

Goulet, F. 1995. Frost heaving of forest tree seedlings: a review. New Forests 9: 67-94.

Grossnickle, S.C. 1993. Shoot water relations and gas exchange of western hemlock and western red cedar seedlings during establishment on a reforestation site. Trees 7: 148-155.

Grossnickle, S.C. 2000. Ecophysiology of northern spruce species: the performance of planted seedlings. Ottawa, ON: National Research Council Research Press. 409 p.

Haase, D.L.; Rose, R.W.; Trobaugh, J. 2006. Field performance of three stock sizes of Douglas-fir container seedlings grown with slow-release fertilizer in the nursery growing medium. New Forests 31: 1-24.

Hainds, M.J. 2003. Determining the correct planting depth for container-grown longleaf pine seedlings. In: Kush, J.S., comp. Longleaf pine: a southern legacy rising from the ashes. Proceedings of the fourth longleaf alliance regional conference. Longleaf Alliance Report No. 6: 66-68.

Hallonborg, U. 1997. Aspects of mechanised tree planting. Uppsala, Sweden: Swedish University of Agricultural Sciences. Doctoral thesis.

Hallsby, G.; Orlander, G. 2004. A comparison of mounding and inverting to establish Norway spruce on podzolic soils in Sweden. Forestry 77: 107-117.

Harstela, P.; Saarinen, V.-M.; Tervo, L.; Kautto, K. 2007. Productivity of planting with M-planter machine. NSFP Nordic Nursery conference, September 5-6, 2007. Suonenjoki, Finland: Finnish Forest Research Institute, Suonenjoki Unit. 2 p.

Heiskanen, J.; Viiri, H. 2005. Effects of mounding on damage by the European pine weevil in planted Norway spruce seedlings. Northern Journal of Applied Forestry 22: 154-161.

Helenius, P. 2005. Effect of thawing regime on growth and mortality of frozen-stored Norway spruce container seedlings planted in cold and warm soil. New Forests 29: 33-41.

Helenius, P.; Luoranen, J.; Rikala, R.; Leinonen, K. 2002. Effect of drought on growth and mortality of actively growing Norway spruce container seedlings planted in summer. Scandinavian Journal of Forest Research 17: 218-224.

Helgerson, O.T. 1989a. Heat damage in tree seedlings and its prevention. New Forests 3: 333-358.

Helgerson, O.T. 1989b. Effects of alternate types of microsite shade on survival of planted Douglas-fir in southwest Oregon. New Forests 3: 327-332.

Henneman, D. 2007. Personal communication. Medford OR: USDI Bureau of Land Management.

Hoag, J.C. 2006. The pot planter: a new attachment for the Waterjet Stinger. Native Plants Journal 7: 100-101.

Hoag, J.C.; Landis, T.D. 2001. Riparian zone restoration: field requirements and nursery opportunities. Native Plants Journal 2: 30-35.

Islam, M.A.; Jacobs, D.F.; Apostol, K.G.; Dumroese, R.K. 2008. Transient physiological responses of planting Douglas-fir seedlings with frozen or thawed root plugs under cool-moist and warm-dry conditions. Canadian Journal of Forest Research 38: 1517-1525.

Jacobs, D.F.; Rose, R.; Haase, D.L. 2003. Incorporating controlled-release fertilizer technology into outplanting. In: Riley, L.E.; Dumroese, R.K.; Landis T.D., tech. coords. National Proceedings, Forest and Conservation Nursery Associations—2002. Proceedings RMRS-P-28. Ogden, UT: USDA Forest Service, Rocky Mountain Research Station: 37-42.

Jacobs, D.F.; Rose, R.; Haase, D.L.; Alzugaray, P.O. 2004. Fertilization at planting inhibits root system development and drought avoidance of Douglas-fir (*Pseudotsuga menziesii*) seedlings. Annals of Forest Science 61: 643-651.

Jacobs, D.F.; Steinbeck, K. 2001. Tree shelters improve survival and growth of planted Engelmann spruce seedlings in southwestern Colorado. Western Journal of Applied Forestry 16: 114-120.

Jeffrey, J.; Horiuchi, B. 2003. Tree planting at Hakalau National Wildlife Refuge—the right tool for the right stock type. Native Plants Journal 4: 30–31.

Johnson, G.R.; Okula, J.P. 2006. Antelope bitterbrush reestablishment: a case study of plant size and browse protection effects. Native Plants Journal 7: 125-133.

Jones, B.; Alm, A.A. 1989. Comparison of planting tools for containerized seedlings: two-year results. Tree Planters' Notes 40(2): 22-24.

Kaye, T.N. 2001. Propagation and population re-establishment for tall bugbane (*Cimicifuga elata*) on the Salem District, BLM. Second year report. Philomath, OR: Institute for Applied Ecology. 12 p.

Khadduri, N. 2008. Personal communication. Olympia, WA: Washington Department of Natural Resources, Webster State Nursery.

Kiiskila, S. 1999. Container stock handling. In: Gertzen, D.; van Steenis, E.; Trotter, D.; Summers, D.; tech. coords. Proceedings of the 1999 Forest Nursery Association of British Columbia. Surrey, BC, Canada: British Columbia Ministry of Forests, Extension Services: 77-80.

Kloetzel, S. 2004. Revegetation and restoration planting tools: an in-the-field perspective. Native Plants Journal 5: 34-42.

Kooistra, C.M.; Bakker, J.D. 2002. Planting frozen conifer seedlings: warming trends and effects on seedling performance. New Forests 23: 225-237.

Kooistra, C.M.; Bakker, J.D. 2005. Frozen-stored conifer container stock can be outplanted without thawing. Native Plants Journal 6: 267-278.

Krumlik, G.J. 1984. Fall-planting in the Vancouver Forest Region. Victoria, BC, Canada: British Columbia Ministry of Forests. Research Rep. 84002-HQ.

Landhäusser, S.; DesRochers, A.; Lieffers, V.J. 2001. A comparison of growth and physiology in *Picea glauca* and *Populus tremuloides* at different soil temperatures. Canadian Journal of Forest Research 31: 1922-1929.

Landis, T.D. 1999. Seedling stock types for outplanting in Alaska. In: Alden J., ed. Stocking standards and reforestation methods for Alaska. University of Alaska Fairbanks, Agricultural and Forestry Experiment Station, Misc. Publication 99-8: 78-84.

Landis, T.D.; Jacobs, D.F. 2008. Hot-planting opens new outplanting windows at high elevations and latitudes. Lincoln, NE: USDA Forest Service. Forest Nursery Notes 28(1): 19-23.

Landis, T.D.; Dumroese; R.K. 2009. Using polymer-coated controlled-release fertilizers in the nursery and after outplanting. Lincoln, NE: USDA Forest Service. Forest Nursery Notes 29(1): 5-12.

Lof, M.; Rydberg, D.; Bolte, A. 2006. Mounding site preparation for forest restoration: survival and short term growth response in *Quercus robur* L. seedlings. Forest Ecology and Management 232: 19-25.

Londo, A.J.; Dicke, S.G. 2006. Measuring survival and planting quality in new pine plantations. Tech. Bull. SREF-FM-001. Athens, GA: University of Georgia, Southern Regional Extension Forestry. 5 p.

Lowman, B. 1999. Tree planting equipment. In: Alden, J. ed. Stocking standards and reforestation methods for Alaska. Misc. Pub. 99-8. Fairbanks, AK: University of Alaska, Agricultural and Forestry Experiment Station: 74-77.

Luoranen, J.; Rikala, R.; Konttinen, K.; Smolander, H. 2005. Extending the planting period of dormant and growing Norway spruce container seedlings to early summer. Silva Fennica 39: 481-496.

Luoranen, J.; Rikala, R.; Smolander, H. 2004. Summer planting of hot-lifted silver birch container seedlings. In: Ciccarese, L.; Lucci, S.; Mattsson, A., eds. Nursery production and stand establishment of broadleaves to promote sustainable forest management; 7-10 May 2001; Rome. Rome, Italy: APAT (Italy's Agency for the Protection of the Environment and for Technical Services): 207-218. http://www.iufro.org/publications/proceedings/ (accessed 23 January 2009).

MacGowan, B.J.; Severeid, L.; Skemp, F. 2004. Control of deer damage with chemical repellents in regenerating hardwood stands. In: Michler, C.H.; Pijut, P.M.; Van Sambeek, J.W.; Coggeshall, M.V.; Seifert, J.; Woeste, K.; Overton, R.; Ponder, F., Jr., eds. Black walnut in a new century, proceedings of the 6th Walnut Council research symposium. Gen. Tech. Rep. NC-243. Lafayette, IN. USDA Forest Service, North Central Research Station: 127-133.

Maki, D.S.; Colombo, S.J. 2001. Early detection of the effects of warm storage on conifer seedlings using physiological tests. Forest Ecology and Management 154: 237-249.

McDonald, P.M.; Fiddler, G.O.; Henry, W.T. 1994. Large mulches and manual release enhance growth of ponderosa pine seedlings. New Forests 8: 169-178.

McDonald, T.P.; Fulton, J.P.; Darr, M.J.; Gallagher, T.V. 2008. Evaluation of a system to spatially monitor hand planting of pine seedlings. Computers and Electronics in Agriculture 64: 173-182.

McKay, H.M.; Gardiner, B.A.; Mason, W.L.; Nelson, D.G.; Hollingsworth, M.K. 1993. The gravitational forces generated by dropping plants and the response of Sitka spruce seedlings to dropping. Canadian Journal of Forest Research 23: 2443–2451.

Meikle, T.W. 2008. Personal communication. Hamilton, MT: Great Bear Restoration.

Miller, D.L.; Brewer, D.W. 1984. Effects of site preparation by burning and dozer scarification on seedling performance. For. Tech. Pap. TP-91-1. Lewiston, ID: Potlatch Corp.

Mitchell, W.K.; Dunsworth, G.; Simpson, D.G.; Vyse, A. 1990. Seedling production and processing: container. In: Lavender, D.P.; Parish, R.; Johnson, C.M.; Montgomery, G.; Vyse, A.; Willis, R.A.; Winston, D. Regenerating British Columbia's forests. Vancouver, BC, Canada: University of British Columbia Press: 235-253.

Moore, J.A.; Fan, Z. 2002. Effect of root-plug incorporated controlled-release fertilizer on two-year growth and survival of planted ponderosa pine seedlings. Western Journal of Applied Forestry 17: 216-219.

Munshower, F.F. 1994. Practical handbook of disturbed land revegetation. Boca Raton, FL: CRC Press. 265 p.

Nelson, J.A. 1984. Elk springs burn seedling survival study—July 1982 to April 1984. Mescalero, NM: Bureau of Indian Affairs, Mescalero Agency. 14 p.

Neumann, R.W.; Landis, T.D. 1995. Benefits and techniques for evaluating outplanting success. In: Landis, T.D.; Cregg, B., tech. coords. National Proceedings, Forest and Conservation Nursery Associations. Gen. Tech. Rep. PNW-GTR-365. Portland, OR: USDA Forest Service, Pacific Northwest Research Station: 36-43.

Nieuwenhuis, M.; Egan, D. 2002. An evaluation and comparison of mechanised and manual tree planting on afforestation and reforestation sites in Ireland. International Journal of Forest Engineering 13: 11-23.

Nilsson, U.; Orlander, G. 1995. Effects of regeneration methods on drought damage to newly planted Norway spruce seedlings. Canadian Journal of Forest Research 25: 790-802.

Orlander, G.; Hallsby, G.; Gemmel, P.; Wilhelmsson, C. 1998. Inverting improves establishment of *Pinus contorta* and *Picea abies*: 10-year results from a site preparation trial in northern Sweden. Scandinavian Journal of Forest Research 13: 160-168.

Page-Dumroese, D.S.; Dumroese, R.K.; Jurgensen, M.F.; Abbott, A.; Hensiek, J.J. 2008. Effect of nursery storage and site preparation techniques on field performance of high-elevation *Pinus contorta* seedlings. Forest Ecology and Management 256: 2065-2072.

Paterson, J.; DeYoe, D.; Millson, S.; Galloway, R. 2001. Handling and planting of seedlings. In: Wagner, R.G.; Colombo, S.J., eds. Regenerating the Canadian forest: principles and practice for Ontario. Markham, ON, Canada: Ontario Ministry of Natural Resources and Fitzhenry & Whiteside Ltd.: 325-341.

Pearce, C. 1990. Monitoring regeneration programs. In: Lavender, D.P.; Parish, R.; Johnson, C.M.; Montgomery, G.; Vyse, A.; Willis, R.A.; Winston, D. Regenerating British Columbia's forests. Vancouver, BC, Canada: University of British Columbia Press: 98-116.

Petersen, G.J. 1982. The effects of artificial shade on seedling survival on western Cascade harsh sites. Tree Planters' Notes 33(1): 20-23.

Peterson, A.; Charton, J. 1999. Advantages and disadvantages of machine planting in south-central Alaska. In: Alden, J. ed. Stocking standards and reforestation methods for Alaska. Misc. Pub. 99-8. Fairbanks, AK: University of Alaska, Agricultural and Forestry Experiment Station: 68-73.

Rose, R. 1992. Seedling handling and planting. In: Hobbs, S.D.; Tesch, S.D.; Owston, P.W.; Stewart, R.E.; Tappeiner, J.C.; Wells, G.E. Reforestation practices in southwestern Oregon and northern California. Corvallis, OR: Oregon State University, Forest Research Laboratory: 328-344.

Rose, R.; Haase, D. 1997. Thawing regimes for freezer-stored container stock. Tree Planters' Notes 48(1&2): 12-18.

Rose, R.; Haase, D.L. 2006. Guide to reforestation in Oregon. Corvallis, OR: Oregon State University, College of Forestry. 48 p.

Rose, R.; Ketchum, J.S. 2002. Interaction of vegetation control and fertilization on conifer species across the Pacific Northwest. Canadian Journal of Forest Research 32: 136-152.

Rose, R.; Rosner, L. 2005. Eighth-year response of Douglas-fir seedlings to area of weed control and herbaceous versus woody weed control. Annals of Forest Science 62: 481-492.

Saarinen, V. 2007. Productivity, quality of work and silvicultural result of mechanized planting. Nordic nursery conference, Sept. 5, 2007. Suonenjoki, Finland: Finnish Forest Research Institute, Suonenjoki Research Station. 13 p. Website: http://www.metla.fi/tapahtumat/2007/nsfptaimitarharetkeily/abstracts/nsfp050907-saarinen.pdf (accessed 16 February 2008).

Sahlen, K.; Goulet, F. 2002. Reduction of frost heaving of Norway spruce and Scots pine seedlings by planting in mounds or in humus. New Forests 24: 175-182.

Scotts Company. 2007. Agriform planting tablets. http://www.scottspro.com/_documents/tech_sheets/H5108_Agriform_20_10_5.pdf (accessed 21 February 2009).

Sharpe, A.L.; Mason, W.L.; Howes, R.E.J. 1990. Early forest performance of roughly handled Sitka spruce and Douglas fir of different plant types. Scottish Forestry 44: 257-265.

Shoulders, E. 1958. Scalping, a practical method of increasing plantation survival. Forest Farmer 17(10): 10-11

Slusher, J.P. 1993. Mechanical tree planters. Pub. G5009. Columbia, MO: University of Missouri-Columbia, Extension Publications. 5 p.

South, D.B. 2008. Personal communication. Auburn, AL: Auburn University, Department of Forestry and Wildlife Sciences.

Saint-Amour, M. 1998. Evaluation of a powered auger for planting large container seedlings. Forest Engineering Research Institute of Canada, Field Note: Silviculture—107. 2 p.

Stein, W.I. 1992. Regeneration surveys and evaluation. In: Hobbs, S.D.; Tesch, S.D.; Owston, P.W.; Stewart, R.E.; Tappeiner, J.C.; Wells, G.E., eds. Reforestation practices in southwestern Oregon and northern California. Corvallis, OR: Oregon State University, Forest Research Laboratory: 346-382.

Steinfeld, D.E.; Riley, S.A.; Wilkinson, K.M.; Landis, T.D.; Riley, L.E. 2007. Roadside revegetation: an integrated approach to establishing native plants. Pub. FHWA-WFL/TD-07-005. Vancouver, WA: Federal Highway Administration, Western Federal Lands High-way Division, Technology Deployment Program. 413 p.

Stroempl, G. 1990. Deeper planting of seedlings and transplants increases plantation survival. Tree Planters' Notes 41(4): 17-21.

Sutherland, B.; Foreman, F.F. 2000. Black spruce and vegetation response to chemical and mechanical site preparation on a boreal mixedwood site. Canadian Journal of Forest Research 30: 1561-1570.

Sutton, R.F. 1993. Mounding site preparation: a review of European and North American experience. New Forests 7: 151-192.

Sutton, R.F.; Weldon, T.P. 1993. Jack pine establishment in Ontario: 5-year comparison of stock types + Bracke scarification, mounding, and chemical site preparation. Forestry Chronicle 69: 545–553.

Swistock, B.R.; Mecum, K.A.; Sharpe, W.E. 1999. Summer temperatures inside ventilated and unventilated brown plastic treeshelters in Pennsylvania. Northern Journal of Applied Forestry 16: 7-10.

Tabbush, P.M. 1986. Rough handling, soil temperature, and root development in outplanted Sitka spruce and Douglas-fir. Canadian Journal of Forest Research 16: 1385-1388.

Talbert, C. 2008. Achieving establishment success the first time. Tree Planters' Notes 53(2): 31-37.

Tan, W.; Blanton, S.; Bielech, J.P. 2008. Summer planting performance of white spruce 1+0 container seedlings affected by nursery short-day treatment. New Forests 35: 187-205.

Taylor, E. 2005. Shift of weather patterns necessitates rethinking of reforestation methods. Texas A&M University, Agricultural Communications. http://agnews.tamu.edu/dailynews/stories/FRSC/May2705a.htm (posted 28 May 2005).

Thomas, D.S. 2008. Hydrogel applied to the root plug of subtropical eucalypt seedlings halves transplant death following planting. Forest Ecology and Management 255: 1305-1314.

Tinus, R.W. 1996. Cold hardiness testing to time lifting and packing of container stock: a case history. Tree Planters' Notes 47(2): 62-67.

Trent, A. 1999. Improved tree-planting tools. Timber Tech Tips 9924-2316-MTDC. Missoula, MT: USDA Forest Service, Technology and Development Program. 6 p.

Troy, T.; Loewenstein, E.; Chappelka, A. 2006. Effect of animal browse protection and fertilizer application on the establishment of planted Nuttall oak seedlings. New Forests 32: 133-143.

Truax, B.; Gagnon, D. 1993. Effects of straw and black plastic mulching on the initial growth and nutrition of butternut, white ash and bur oak. Forest Ecology and Management 57: 17-27.

(USDA Forest Service) U.S. Department of Agriculture. 2002. Silvicultural practices handbook (2409.17), Chapter 2—reforestation. Missoula, MT: USDA Forest Service. 106 p.

van den Driessche, R. 1987. Importance of current photosynthate to new root growth in planted conifer seedlings. Canadian Journal of Forest Research 17: 776-782.

van den Driessche, R. 1988. Response of Douglas-fir (*Pseudotsuga menziesii* (Mirb.) Franco) to some different fertilizers applied at planting. New Forests 2: 89-110.

White, J.J. 1990. Nursery stock root systems and tree establishment: a literature review. Occ. Pap. 20. Edinburgh, United Kingdom: Forestry Commission. 43 p.

Zalasky, H. 1983. Field storage of containerized conifer seedlings. Forest Management Note 20. Edmonton, AB, Canada: Northern Forest Research Centre. 3 p.

www.ingramcontent.com/pod-product-compliance
Ingram Content Group UK Ltd.
Pitfield, Milton Keynes, MK11 3LW, UK
UKHW051240180426
11947UKWH00013B/868